SYMMETRIES
— OF —
ISLAMIC GEOMETRICAL PATTERNS

SYMMETRIES
—— OF ——
ISLAMIC GEOMETRICAL PATTERNS

by
Syed Jan Abas and **Amer Shaker Salman**
School of Mathematics, Univ. of Wales
Bangor

With forewords by
Ahmed Moustafa
Celebrated Arabic Calligrapher

and

Sir Michael Atiyah O.M, *F.R.S,P.R.S.*
President of the Royal Society
Master of Trinity College Cambridge

World Scientific
Singapore • New Jersey • London • Hong Kong

Published by

World Scientific Publishing Co. Pte. Ltd.
P O Box 128, Farrer Road, Singapore 912805
USA office: Suite 1B, 1060 Main Street, River Edge, NJ 07661
UK office: 57 Shelton Street, Covent Garden, London WC2H 9HE

British Library Cataloguing-in-Publication Data
A catalogue record for this book is available from the British Library.

First published 1995
First reprint 1998

SYMMETRIES OF ISLAMIC GEOMETRICAL PATTERNS

Copyright © 1995 by World Scientific Publishing Co. Pte. Ltd.

All rights reserved. This book, or parts thereof, may not be reproduced in any form or by any means, electronic or mechanical, including photocopying, recording or any information storage and retrieval system now known or to be invented, without written permission from the Publisher.

For photocopying of material in this volume, please pay a copying fee through the Copyright Clearance Center, Inc., 222 Rosewood Drive, Danvers, MA 01923, USA. In this case permission to photocopy is not required from the publisher.

ISBN 981-02-1704-8

Printed in Singapore by Eurasia Press Pte Ltd

PREFACE: ABOUT THE BOOK

The artist and the mathematician in Arab civilization have become one. And I mean quite literally.

Jacob Bronowski

WHAT IS THE BOOK ABOUT?

This book is about Islamic geometrical patterns.[1] It has educational, aesthetic, cultural, and practical purposes.

The central purpose of the book is to bring to the attention of the world in general, and the people of Islamic countries in particular, the potential of Islamic symmetric patterns for providing a unified experience of science and art in the context of geometrical and mathematical education. Such experience has enormous value, not only for mathematicians, but also for artists, designers, computer scientists, physicists, chemists, crystallographers, art historians, archaeologists and others. Beyond the needs of education, the experience of science and art in unison is satisfying to the aesthetic and cultural needs of intelligent beings everywhere.

The potential of Islamic art for offering such experience arises from the fact that it relies primarily on geometry and on explorations of pattern and symmetry. The dominion of geometry, as the wisest through the ages have proclaimed, is supreme. The harmony of the Universe can only truly be experienced in the purest perfections of geometrical form.

Geometrical pattern and symmetry, which comprise the main visible body of Islamic art, lead naturally to abstract notions of pattern symmetry.

---

[1] It is not easy to give a precise definition of what is meant by an *Islamic* pattern. We shall discuss the matter in the next chapter.

These are the two most profound and wide ranging notions that the human brain can conceive. They unite science, art, and nature as nothing else does. Pattern and symmetry can be as absorbing and meaningful to a child in the kindergarten as to a sophisticated high energy physicist who theorizes about the building blocks of the Universe.

Although all human cultures, from the earliest of times, have explored pattern and symmetry, it was in the Islamic civilization, around 10th century, that the activity truly began to blossom. The art reached its zenith in the mid 14th century, resulting in such magnificent creations as are to be found in the Nasrid Palace of Alhambra, in Granada, Spain.

Then, as the Islamic civilization and learning began to decline, so this activity lost its vigor and spent itself. It is true that in some places, particularly in Morocco, the ancient skills are alive and flourishing but no real major innovation can be observed. Our purpose is also to encourage a renaissance in the art of pattern and symmetry through the use of computer graphics and more advanced mathematics than simple two-dimensional Euclidean geometry. This combination offers all sorts of new and exciting possibilities for the revival of the Islamic tradition and its future growth. Islamic patterns on polyhedral surfaces, patterns on algebraic surfaces, non-periodic patterns, color symmetry, hyperbolic symmetry, patterns in virtual reality, and the use of non-linear grids, are some examples.

We begin chapter 1 by introducing the reader to Islamic patterns and showing their key geometrical structures.

In chapter 2 we discuss the immensity of the notions of pattern and symmetry in general terms and show how Islamic patterns are particularly suited for artistic celebrations of symmetrical forms that occur at molecular levels of reality.

Symmetric geometrical patterns contain embedded within them the abstract mathematical notions of invariance and group. These currently reside at the core of scientific thought and provide us with powerful ways of understanding the Universe. Without making any technical demands on the part of the reader beyond school level mathematics, we explain in chapter 3 the meaning and significance of the abstract notions of symmetry, invariance and group. We discuss how invariance unites not only science and art but also religion and philosophy in a very general way. We point out the supreme importance of the question *What abides?* in all the major branches of human endeavor.

Chapter 4 of the book builds on chapter 3 to explain how patterns are classified and how they can be recognized from their symmetry properties. It also gives simple algorithms for the construction of the 17 types of repeat patterns which can arise in two dimensions.

Over the last few years the authors have studied several hundred Islamic geometrical patterns using computer graphics [1,2,42]. Chapter 5 results from this study, where we offer a collection of 250 patterns. The patterns in this collection are presented in such a way as to encourage their appreciation from scientific as well as artistic points of view.

Chapter 5 also caters for the practical purpose of the book. The same analysis which has been used by us to classify the patterns has also been utilized to extract numerical data for use with computer graphics.

This data[2] may be utilized not only for re-creating the original patterns, but can serve as templates for producing an unlimited number of new variations. Using computer aided design and manufacturing (CAD, CAM), it may be exploited to create classical as well as new designs on a variety of materials such as wood, ceramics, concrete, textile, glass, precious metals and others. Furthermore, the same two-dimensional data can be used to extend Islamic patterns to three-dimensional surfaces and in a variety of other ways.

Finally, as another by-product of symmetry analysis in our studies of Islamic patterns, we present in chapter 5 a characteristic feature of the Islamic culture in its choice of symmetry type. The very surprising discovery that certain symmetry types are preferred and intuitively recognized as being *right* by each culture is a recent one and has emerged from research in archaeology [51]. Our book identifies this feature of Islamic culture.

HOW DOES THE BOOK DIFFER FROM OTHERS?

Previous books on the subject of Islamic patterns have set out to offer large numbers of ad hoc constructions for large collections of Islamic patterns. With the exception of Lalvani [23,24], the authors have viewed Islamic geometrical patterns either as mere surface decorations, or have related them only to religious, mystical, astrological, and other esoteric themes. Their relation to modern science, to more advanced mathematical topics beyond simple geometrical constructions, and their general educational value has not been greatly emphasized. Neither have their constructions been presented in a form suitable for innovation and advancement.

Only a few isolated individuals in the West have commented on the greater scientific merits of Islamic art. Jacob Bronowski was one in this rare category who in his book *The Ascent of Man* wrote [7]:

---

[2]The data is not included in this book. For the explanation, see chapter 5.

> *The artist and the mathematician in Arab civilization have become one. And I mean quite literally. These patterns represent a high point of the Arab exploration of the subtleties and symmetries of space itself ...*
>
> *Thinking about these forms of patterns, exhausting in practice the possibilities of the symmetries of space, at least in two dimensions, was the great achievement of Arab mathematics.*

Emil and Milota Makovicky and Hary Bixler are some other authors of whom we are aware who have tried to point out wider educational[3] and cultural significances of Islamic patterns. The Makovickys in a paper entitled *Arabic geometrical patterns — a treasury for crystallographic teaching* [26] extolled the potential of Islamic patterns for the teaching of group theory and crystallography. Harry Bixler, in his Ph.D. thesis entitled *A Group-theoretic Analysis of Symmetry in two-dimensional Patterns from Islamic Art*, submitted to New York University [5], set out to show how Islamic art can provide a two way bridge between science and art. He concluded:

> *the world of two-dimensional Islamic art affords the opportunity par excellence of illustrating structural similarities between the world of art and mathematics.*

Our book sets out to place the study of Islamic geometrical patterns in the wider framework of the study of pattern and symmetry. It is intended to encourage innovation. The author[4] endorses Jean Piaget's dictum:

> *The principal goal of education is to create men and women who are capable of doing new things, not simply of repeating what other generations have done — men and women who are creative, inventive and discoverers.*

---

[3] At the school level, many mathematics teachers in the West have reported their creative usage of Islamic patterns, for example [35].

[4] The text in this book and the ideas expressed are entirely due to the first author SJA. The second named author has drawn most of the black and white diagrams. Also, the analysis of symmetry and the extraction of template motifs in chapter 5 are substantially the work of the second author ASS. Throughout the book, the term *The author* is used as a short form for *The first author SJA*.

THE INTENDED AUDIENCE

The book is aimed at the general audience and makes no specialist demands on the part of the reader beyond school level mathematics. We hope that it will appeal to scientists, artists, educationalists, designers, humanists, and intelligent lay persons everywhere. It ought to appeal specially to teachers of mathematics.

Whereas the geometrical patterns of Islamic art can serve artistic, scientific, and educational needs in all cultures, they are of course of special value for the educators in Islamic cultures. They offer something truly excellent and of universal value from the past which may be conjoined with the learning of today.

At its height, the Islamic civilization contributed extensively to preserving and promoting the intellectual works of humankind. It safeguarded and developed the geometry of the Greeks while Europe was slumbering in the Dark Ages. It took the number system from India and transplanted it in Europe. It invented Algebra and formulated the concept of Algorithm.

Today, the language of mathematics built on Arabic numerals, geometry, Algebra, and Algorithms is the most powerful language there is. It is the language of those who design computers, build spaceships, forecast the weather, and speculate about the beginnings of the Universe. For the educators in Islamic countries, the geometrical patterns of Islamic art can serve not only to illuminate historical achievements, but more importantly, they can be used to initiate learning in today's disciplines. They deserve to be encouraged, broadened, and fostered in the modern context. They constitute ideal landmarks for starting today's journeys into numerals, geometries, algebras, and algorithms.

For these reasons it is hoped that the book will have a special appeal for audiences in the Islamic world. We shall be gratified if the book manages to encourage the forging of links between the past and the present and the explorations of new patterns, new symmetries, and new colors which bring new credit to the old traditions of Islamic art.

ACKNOWLEDGMENTS

Although we have consulted numerous references and taken several study trips, our knowledge of Islamic patterns has come substantially from examining the pioneering collection published by Bourgoin [6] and from the books by Critchlow [9], El-Said and Parman [13], and Wade [50]. We pay tribute to the scholarship and insight of these authors. Our work would not have been possible without their earlier contributions.

We have also been fortunate enough to receive encouragement, support, collaboration, and kindness from many talented individuals and helpful organizations. It is a privilege for us to acknowledge their generosity and willingness to give us their valuable time. In particular, we wish to express our gratitude to the following:

Professor Ronald Brown, the Head of the School of Mathematics at the University of Wales Bangor and Mr. Terry Hewitt, the Director of the Manchester University Computer Graphics Unit for their wide ranging support in our research and writing activity over many years.

Mr. Alistair Duncan, Professor Myron Evans, Mr. Najib Gedal, Dr. Cliff Pickover, and Professor Brian Rudall for offering us encouragement and kind words on examining our initial proposal for the book.

Mr. Chonglin Chen, Mr. Greg Edwards, Mr. Dahabi Idrissi, Professor Akhlesh Lakhtakia, Professor Haresh Lalvani, Dr. Ramez Ghazoul, Professor R. Messier, Professor Rahim Mirbahar, Professor Rafael Pérez-Gómez, Mr. Samir Shakir Mahmood; Artizana, The Gallery, Prestbury, Cheshire, England; The Islamic Design Company Ltd. London, Lappec UK, Samir Design Ltd. London, Magreb Arab Press London, The Moroccan Tourist office London, and the Turkish Tourist Office London for donating photographs. We have only been able to include a very small selection but are grateful to all for their willingness to help.

The President of the University of Istanbul Library, for supplying and giving permission to use the slide shown in the Color Plate 1(a). This comes from the copyright manuscript MS No. FY1404 held at the Library.

Mr. Tim Lambert, Mr. Martin Owen, and Mr. Kevin Spencer for help with the production of photographs and images.

Mr. Owain Roberts for doing an initial water color of Mahan for the color plate 7.

Dr. Kouichi Honda for giving permission to use his calligraphy on the Dedication pages and with our computer graphics software, for the color plate 15.

Mr. Aziz Ahmed for kindness and help throughout and for inscribing Urdu Calligraphy on the Dedication Page by the author.

A special Thank you to Mr. Greg Edwards of Silicon Graphics, Reading, U.K. for collaboration in producing 3-D computer graphics for the color plates 12, 15 and 16. This work was done using Silicon Graphics' 3-D modeling software the *Inventor*.

My colleagues Dr. Neil Rymer and Mr. Terry Williams for reading the manuscript and making suggestions for improvement. This is the second time Terry Williams has given his valuable time to reading and criticizing a book by the author. He is most grateful. Any remaining errors and omissions are our own.

Finally we wish to thank our distinguished contributors Dr. Ahmed Mostafa and Professor Sir Michael Atiyah, F.R.S., P.R.S., O.M. for their kindness in agreeing to offer their views on pattern, symmetry, unity, and Islamic art as short forewords.

Syed Jan Abas
Amer Shaker Salman
April 1994

I dedicate this book humbly to
my mother in the sacred memory of
her last words:

and to those who conceived and built
The Alhambra.

S.J.Abas
April 1994

صَدَقَ اللهُ الْعَظِيم

وَبِالْوَالِدَيْنِ إِحْسَانًا

To my parents

and

To Sarah

A.S.Salman
April 1994

FOREWORD

Symmetry is one of the most important and pervasive principles in Mathematics, particularly in its Geometrical form. Here, mathematics combines with art and exhibits clearly its aesthetic appeal. Islamic patterns provide a marvellous illustration of symmetry and Drs. Abas and Salman perform a useful service by taking this as their theme and blending it with ideas on computer graphics.

Michael Atiyah
Trinity College, Cambridge

FOREWORD

Despite the assumption of many Westerners, Islam has no difficulty in reconciling science and religion. On the contrary, the Qu'ran clearly urges us to deepen our spirituality through a greater understanding of the visible world, and the Prophet Mohammed (P.b.u.H.) emphasised that the search for knowledge is "strict duty" of every Muslim. There is no inconsistency here. This approach inevitably leads, as the Qu'ran points out, to the recognition of a greater reality beyond the limits of our perception — "Of all his servants, only such as are endowed with (innate) knowledge stand (truly) in awe of God; (for they alone comprehend that,) verily, God is Almighty, much-forgiving"; Verse 28, Surah 35.

It was this insistence on objective investigation that brought about, among other things, the study of Ancient Greece and, in a relatively short time, the translation into Arabic of its science, its art, and its philosophy.

By the end of the 9th century AD, Muslim scholars, who had gone about their mission with what Bronowski (The Ascent of Man) described as "kleptomanic zest", were describing geometry as a reflection of Divine creation and an illustration of Verse 49, Surah 54 of the Qu'ran: "Behold, everything have We created in due measure and proportion."

It is easy to image the excitement and recognition of those muslim scholars, those I call the scientists of the Art of Islam, reading Plato's conclusion: (Geometry) has the effect of making it easier to see the form of the good. And that, we say, is the tendency of everything which compels the mind to turn to the region of ultimate blessedness which it must spurn no effect to see.

My own research into the work of the distinguished 10th century geometer Ibn Muqlah showed clearly how his theory of Proportioned Script was rooted in the Pythagorean theorem which constitutes the symmetry of plane space and which I believe mirrors the harmony of Nature. The encounter by Muslim scholars with the body of Greek science and intellectual thought informed Ibn Muqlah's Theorem which became central to the visual harmony of Islamic Art and Architecture. It also provided the proportional measure governing absolutely the surface area of each individual Arabic letter shape and its visual relation to the rest of the Alphabet.

Although the Babylonians and the Ancient Egyptians were certainly aware of the symmetry of space and its practical implications, it was not until 550BC that Pythagoras, in his theorem, raised the notion from the merely empirical to a standard of proof, utilised to such effect by Ibn Muqlah some 14 centuries later.

So the not uncommon notion that religious interdiction prevented Islamic artists from depicting images from life and forced them into abstraction is unfounded and simply wrong. The fact is that Islamic civilisation found in Geometry a verification of its beliefs and a system which enabled Artists to extend their creativity according to a law which is part of nature itself and offers an endless field of exploration. The direction and subsequent efflorescence of mainly abstract Islamic Art was the result neither of compulsion nor obligation but of choice. Plato had maintained the same choice some 1,000 years before the advent of Islam: "There is a faculty in the mind of each of us which these (geometrical) studies purify and rekindle after it has been ruined and blinded by other pursuits, though it is more worth preserving than any eye since it is the only organ by which we perceive the truth."

Regrettably few publications on Islamic Art adequately address its governing principles which tend to be lost in a flattening and obscure vocabulary. This book by Dr. Abas is, therefore, a most welcome and important contribution to the subject, recognising as it does the place of Geometry as the bedrock of Islamic Art and the essential understanding that its patterns, far from being mere abstract niceties, are nothing less than visual homage to God, expressed in His own immutable goemetrical discipline.

Ahmed Moustafa
London
26 Dhul-hija 1414
5 June 1994

CONTENTS

Preface: About the Book	v
Acknowledgments	xi
Dedication by S. J. Abas	xiii
Dedication by A. S. Salman	xv
Foreword by Michael Atiyah	xvii
Foreword by Ahmed Moustafa	xix
1. Islamic Patterns and Their Geometrical Structures	1
1.1. Islamic Patterns: An Introduction	2
1.1.1. Types of Patterns	2
1.1.2. Recognizable Characteristics	4
1.1.3. A Definition	6
1.1.4. Turning to Geometry	8
1.1.5. Related Literature	12
1.2. Geometrical Strategies and Structures	14
1.2.1. Khatem Sulemani: The Most Basic Shape	14
1.2.2. Variations on Khatem Sulemani	16
1.2.3. Harmonious Proportions: The Secret of Beauty	18
1.2.4. A Complex Pattern	19
1.2.5. Grids and Circles	21
1.2.6. The Zalij Approach to Islamic Patterns	24
1.3. Concluding Remarks	27
2. In Praise of Pattern, Symmetry, Unity & Islamic Art	29
2.1. In Praise of Pattern	30
2.1.1. In Praise of Symmetry	32
2.2. In Praise of Unity	35
2.3. In Praise of Islamic Art	36
2.3.1. Islamic Patterns and Atomic and Molecular Structures	37
2.3.2. A Celebration of Unity	39
2.4. Concluding Remarks	43

3. The Gateway from Islamic Patterns to Invariance and Groups	45
3.1. An Example of Symmetries of an Islamic Pattern	46
3.2. The Key Property of Symmetric Objects	54
3.2.1. Finite Symmetric Objects with Centres of Rotations and Lines of Mirror Reflections	54
3.2.2. Infinite Symmetric Objects with Translations and Lines of Glide Reflections	56
3.2.3. Definition of Symmetry for Geometrical Objects	57
3.2.4. Symmetry Group of a Geometric Object	58
3.2.5. The Four Special Properties of a Symmetry Group	59
3.3. The Symmetry Group of an Islamic Pattern	61
3.3.1. The Translation Symmetries	62
3.3.2. The Rotation Symmetries	63
3.3.3. The Mirror Reflection Symmetries	64
3.3.4. The Glide Reflection Symmetries	65
3.3.5. Symmetries Depicted in a Unit Cell	65
3.4. Symmetry and Groups in General	66
3.5. The Two Grand Questions and One that is Grandest	69
3.6. Concluding Remarks	71
4. Classification, Identification and Construction of the Seventeen Types of Two-Dimensional Periodic Patterns	73
4.1. Classification of Patterns	74
4.1.1. The Five Net Types	75
4.1.2. The Seventeen Pattern Types	76
4.1.3. The International Crystallographic Notation	77
4.2. Examples of the Seventeen Types of Patterns from Islamic Art	79
4.3. Identifying the Seventeen Pattern Types	108
4.4. Tile-Based Algorithms for the Seventeen Pattern Types	114
4.4.1. The Notation	114
4.4.2. The Procedure	115
4.4.3. An Example	115
4.5. Concluding Remarks	116
5. Islamic Patterns and Their Symmetries	135
5.0.1. Related Works	136
5.0.2. Preferred Symmetry Types in the Islamic Culture	138
5.1. Concluding Remarks	139
References	389
Index	395

Colour Plates

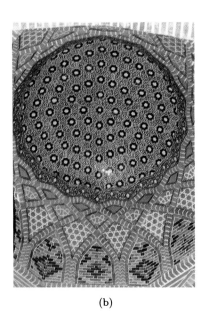

(b)

(a)

Plate 1. (a) Taqi-al-Din and fellow astronomers at work in an observatory built at Istanbul in the 16th century. (b) An interior view of a dome in the Shahjahan Mosque, Tatha, Pakistan; 17th century.

(b)

(a)

Plate 2. (a) The author engrossed in a spectacular Islamic pattern; from the Marinid era (1258-1358 A.D.), now in Musee Al Kasabah, Tangier, Morocco. (b) Patterns in the Alcazar, Seville, Spain.

Plate 3. Four examples of patterns from the Alhambra, Granada, Spain. It was here that the Islamic art of pattern and symmetry reached its greatest height in the 14th century.

Plate 4. To do full justice to them, the patterns must be seen in their environment under the combined effects of proportions, color, light, water and so on. Here, the pattern in plate 3(a) is shown in more of its setting.

(b)

(a)

Plate 5. (a) Tillya-Kari Medersa, Samarkand, Uzbekistan; 17th century. (b) Kalyan Minaret, Bukhara, Uzbekistan; 12th century. A beautiful example of ornamental brick work.

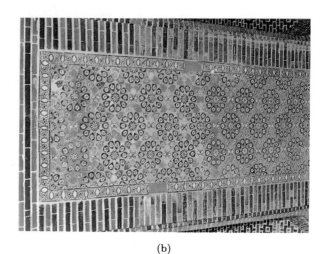

(b)

(a)

Plate 6. (a) Patterns in wood; from a 16th-18th century mansion in Cairo, Egypt. Bought and refurbished in 1930 by a Major Gayer-Anderson, hence, now known as the Gayer-Anderson House. (b) From the Shahjahan Mosque, Tatha, Pakistan; 17th century.

Plate 7. A computer 'painting' by the author based on a watercolor sketch by Owain Roberts, showing the shrine of Sufi Shah Nimatullah, Mahan, Iran; early 17th century.

Plate 8. **Still life of Kur'anic Solids:** by Ahmed Moustafa (1987); oil and watercolor on hand-made paper. The calligraphy on the faces of the solids repeats the verse LIV-49 from the Koran *Behold, everything have we created in due measure and proportion* whilst the floor utilizes the verse III-2 *Allah, there is no deity save him, the Ever-Living the Self-Subsistent Fount of All Being.*

Plate 9. (a) A master zaliji cutting zalij tiles on the site of a villa in the outskirts of Tangier, Morocco. (b) A wood carver in Monestir, Tunisia.

Plate 10. A fountain outside the Mohammed V mausoleum, Rabat, Morocco.

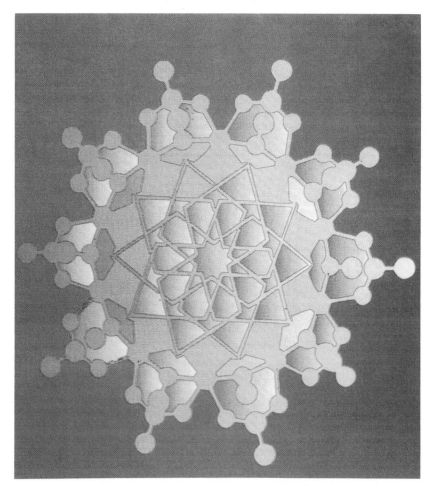

Plate 11. **The Islamic Ferric Wheel:**, a pattern by the author based on the structure of the molecule **Ferric Wheel** synthesized by S. J. Lippard and K. L. Taft of the Massachusetts Institute of Technology. The pattern was designed to celebrate the unity of science and art in Islamic design.

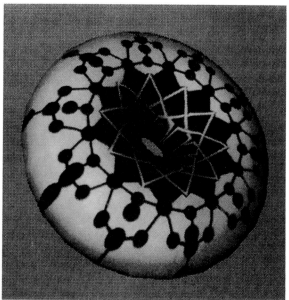

Plate 12. Another version of the Islamic Ferric Wheel covering 3-D surfaces.

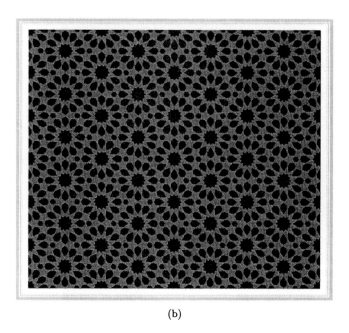

(b)

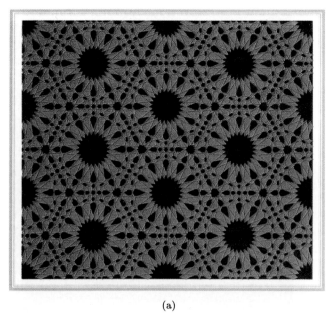

(a)

Plate 13. (a, b) Typical outputs from a program written by the author at the Manchester University Computer Graphics unit to implement the algorithms given in chapter 4.

Plate 14. Examples of Islamic wallpaper based on traditional designs (courtesy Islamic Design Company, Ltd., London).

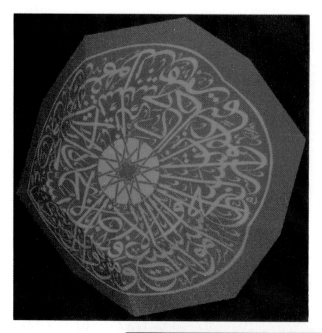

Plate 15

Plate 16

Plates 15, 16. Islamic patterns and Arabic calligraphy on 3-D surfaces. Plate 15 utilizes a calligraphy from the Koran by Professor Koichi Honda of Tokyo University. The calligraphy in plate 15 is based on one inscribed on the dome of the mosque of Hagia Sophia in Istanbul, Turkey; 16th century.

Chapter 1

ISLAMIC PATTERNS AND THEIR GEOMETRICAL STRUCTURES

Allahun Jamilun Ya Habul Jamal
God is beautiful and loves beauty.

From the sayings of Prophet Mohammed

But how shall we define the Infinite?
How shall we fix each fresh and varying phase
That flits for aye across our baffled sight,
And makes us faint and giddy as we gaze?

The lines are from a translation by Edward Palmer [37] of the works of the Turkish mystic saint Rumi (1207–73), the founder of the Mevlevi order of whirling dervishes. Although they were not composed specially for the purpose, they could be utilized to encapsulate the aspirations of Islamic art. How to define that which is infinite and perfectly beautiful? How to capture for transient mortals the essence of the eternal God? How to create images of a perfect paradise on a very imperfect earth?

The aspiration to capture the essence of God is of course common to sacred art of all cultures. But unlike art in other cultures, Islamic art chose mostly to shun anthropomorphic forms.[1]

It focused attention instead on geometry, pattern, symmetry, and on the celebration of sacred and inspiring words in the form of Quranic calligraphy. The answer to Rumi's question, given by Islamic art is that:

> *You shall define the infinite by creating a beautifully ordered symmetric structure which repeats itself for ever and ever and is bejewelled with divine words!*

Of course, it is not being claimed that geometric patterns and schemes are an exclusive prerogative of Islamic art. Far from it. They have been, and continue to be, explored in all cultures. Some, such as the Romans, the Chinese, and the Celts, have employed them extensively in their artistic creations. But nowhere has there been such a concentrated effort to seek out beauty, harmony, and order in elegantly repeating geometrical forms.

In this chapter we shall first introduce the reader to Islamic patterns and point out their easily recognizable characteristics. We shall then offer a definition of Islamic patterns in such a way as to permit their development to continue in the future, and not restrict them to their medieval origins. Next, we shall highlight the intellectual, psychological, and practical reasons which led Islamic cultures to seek inspiration in geometry. We shall follow this with a brief summary of related literature. The final section of this chapter will give insight into the geometrical strategies and structures that underlie Islamic patterns.

1.1 ISLAMIC PATTERNS: AN INTRODUCTION

1.1.1 Types of Patterns

Geometric patterns occur in rich profusion throughout Islamic cultures. They are found on a diversity of materials — tiles, bricks, wood, brass, paper, plaster, glass, and on many types of objects. They occur on carpets, manuscripts and wooden carvings — particularly doors, screens and pulpits or *mimbars*. They can be seen most prominently on architectural surfaces.

---

[1] It would be false to assume that Islamic art is entirely non-figurative. There exists a substantial corpus of figurative work, particularly in the form of miniatures.

1.1. Islamic Patterns: An Introduction

Leaving aside the motifs on carpets and the floral and the stylized floral designs, Islamic patterns come in three distinct geometrical flavors.

One instantly recognizable class is that of the rectangular Kufic patterns. These employ simple rectangles and squares to create calligraphic designs in a stylized form of the Arabic script. Such patterns are employed mostly on architectural surfaces to add dignity and solemnity. An example of a rectangular Kufic pattern is shown in Fig. 1.1. The design inscribes *Mohammed* — the name of the Prophet in stylized Arabic.

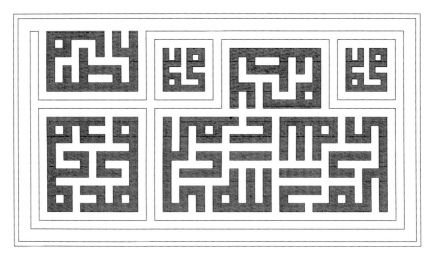

Figure 1.1: A Rectangular Kufic Pattern based on a wall decoration in the Bibi Khanum Mosque, Samarkand, Uzbekistan, early 15th century.

Another distinct pattern type perfected in Islamic art is the arabesque. This comprises curvilinear elements resembling leafed and floral forms. In such patterns spiral forms intertwine, undulate, and coalesce continuously. The sense of periodicity and rhythm is very noticeable in these patterns. The arabesque is an abstraction from the earlier leaf-scroll motifs. An example of arabesque is shown in Photo 1.1.

The largest class of Islamic patterns employs complex polygons and, less frequently, regions bounded by circular arcs. They are *space filling* patterns in which, as we shall see later, the design in a basic cell repeats itself over and over again. In this book, we are only concerned with this third class of pattern. The book is full of examples of such patterns. Chapter 5 contains 250 of them in black and white. A representative range in color is shown in the color Plates, which the reader is now invited to peruse if he or she has not already done so.

4   Chapter 1. Islamic Patterns and Their Geometrical Structures

Photo 1.1: An Arabesque design based on a tile from 15th century Iran, from the British Museum London.

1.1.2 Recognizable Characteristics

Let us now point out some easily recognizable characteristics of Islamic patterns. On examining the color plates, the reader will see that the most striking characteristic of Islamic geometrical patterns is the prominence of symmetric shapes which resemble 'stars' and 'constellations'. Although there exist Islamic patterns with no star shapes in them, they are relatively few and not very sophisticated or complex. Star shapes with six, eight, ten, twelve, and sixteen rays are the ones that occur most frequently, but star shapes with other numbers of rays, particularly in multiples of eight up to ninety six, can be found. This is not a mere decorative exercise. There are important psychological and historical reasons for this which we shall discuss later.

Plates 2(a), 3(a) and 3(b) show another easily recognized feature. The rectilinear elements forming the pattern are often interlaced. This reflects the tent-dwelling, carpet-weaving origins, and experiences of the Muslim populations. The Arabs, the Persians, the Turks from central Asia, the Mongols, and the Berbers have all been tent dwellers. All of them have long histories of carpet weaving where *interlacing is all*. Carpets from the Islamic world continue to be the most sought after today and one is

1.1. Islamic Patterns: An Introduction

not surprised to see the prominence and the refinement of interlacing in the patterns of Islam. Another superb interlacing example is shown in Photo 1.2 which happens to be one of the earliest examples of a fully developed Islamic pattern.

Photo 1.2: Mosaic tile decorations on the interior of the dome of Karatay Madrasah, Konya, Turkey, mid 13th century.

Notice the presence of calligraphy in Plates 2(a), 3(a) and 3(d) and also the curvilinear elements above the patterns in the plates 2(b), 3(a) and 3(d). It is this blending of rectilinear and curvilinear elements with Arabic calligraphy which gives Islamic architectural decoration its most characteristic feature. For the Muslim, the verses from the Quran represent the visual body of the divine word and their presence heightens the sacred and the inspirational elements.

Just as no other civilization has valued pattern and symmetry as deeply as the Islamic civilization, so it is true that no other has revered the sacredness of *the Word* to such an extent. Calligraphy is the jewel in the crown of Islamic art. A magnificent example of a recent calligraphic work by Ahmed Moustafa, which celebrates the spiritual element in geometry, is shown in Plate 8.

Two other characteristic features, which may not be obvious from seeing isolated portions, have to do with *flow* and *unboundedness*. These may be easier to appreciate in Plates 13(a) and (b). As mentioned earlier, in the Islamic patterns of the kind we are discussing, the geometry employed is based on the construction of a repeating cell. By copying the cell, the pattern can be replicated indefinitely to fill as much space as desired. An allied feature is that there is no natural point of focus for the eye. As one looks at an expanse of pattern, the eye 'flows' continuously following the lines and seeing a variety of intricate structures and relationships. This feature is not prominent in the geometric designs of other cultures, such as the Romans.

1.1.3 A Definition

So far, we have used the term *Islamic* pattern without giving a definition of what exactly we mean by *Islamic*. Had we been interested in Islamic patterns as entities only of the past, we might have been excused this omission. But it is our wish that the creation of Islamic patterns should continue into the future. The treasury of Islamic patterns should be offered as a gift of the Islamic civilization for the enrichment of all cultures. We, therefore, feel obliged to address the issue.

The adjective *Islamic* qualifying the noun *pattern* has never implied that the pattern was sanctified in some way by the religion of Islam, or was necessarily invented by a Muslim. Extensive populations of Christians, Jews, Hindus, and others have existed in the Islamic world and have contributed to Islamic art. It is known, for example, that some fine Islamic art objects in the form of marble screens or *Jalis* in some of the Mogul architecture in India are the creations of Hindu craftsmen.

To get round this problem, art scholars have often taken the adjective *Islamic*[2] in the context of art to mean — *"the art produced by a culture or civilization in which the majority of the population, or at least the ruling element, profess the faith of Islam. The artist who actually produced a work of Islamic art may or may not be a Muslim"*.

For us this interpretation of *Islamic* will not do. Figure 1.2 shows a pattern which was made by the author in the U.K., where the majority do not profess the faith of Islam. The construction of the pattern employs a totally different type of geometrical technique[3] from the ones that have

---

[2]See, for example, page 2 of *The Formation of Islamic art* by Oleg Grabar, Yale University Press (1987).

[3]This is an Islamic Penrose pattern. A book by the author on such patterns is being prepared.

1.1. Islamic Patterns: An Introduction

Figure 1.2: A new Islamic pattern.

been used traditionally. Yet anyone who is at all familiar with Islamic culture will instantly recognize it as an *Islamic* pattern.

We will define an Islamic pattern in the following way:

Definition: An Islamic pattern is one which satisfies one or more of the following criteria:

1. The pattern is transcribed with Arabic Calligraphy from the Quran.

2. The pattern was invented between 900 A.D. and 1500 A.D. and was used to decorate architectural surfaces or other works of art for Muslims, in a culture where the majority of the population, or at least the ruling element, professed the faith of Islam.

3. The pattern is derived from one or more patterns which satisfy criterion 2 and is such that the characteristic shapes from the original (or originals) are recognizable.

Let us explain briefly the motives behind our definition. Criterion 1 should be obvious. Color plate 15 was produced in England and uses calligraphy done in Japan. By virtue of the fact that it inscribes verses from the Quran, it is Islamic.

In criterion 2, the choice of the dates arises from the fact that Islamic art only began to acquire a distinct personality in the 10th century. As far as explorations of symmetry are concerned, all the major innovations occurred from the 12th to the 14th century. Everything else after that can be seen as a straight forward derivation and will satisfy criterion 3. By extending the range of dates as far as 1500 A.D. we are being ultra cautious, to cover the period of invention.

The phrase *characteristic shapes* in criterion 3 is vague and we accept this. The question of whether or not a pattern satisfies this criterion must be left to an observer who is familiar with Islamic patterns of category 2. This is not as arbitrary as it appears. Anyone who is familiar with the face of the President of the United States, say, can instantly recognize a cartoon depicting him, even though the features may have been grossly distorted and transformed. The human brain has astonishing powers of recognizing characteristic shapes even though we may not be able to give a precise explanation. Islamic patterns contain a variety of recognizable polygonal characteristic shapes. Indeed, currently, the practice of the art of Islamic patterns in its most developed form, namely, the art of zalij in Morocco, is based on the recognition of characteristic shapes. We shall later describe the art of zalij and point out several characteristic shapes used in it.

With our definition the pattern in Fig. 1.2 qualifies as an Islamic pattern. It is in fact based on a mid 14th century pattern from the Friday Mosque in Yazd, Iran. It is different from the original on which it is based, but contains its prominent and recognizable characteristic Islamic shapes. The fact that it was invented in 1994 in the U.K. should not prevent it from being called an *Islamic* pattern.

Comment: We have constructed this definition to encourage the creation of new Islamic patterns in all parts of the world. This book is concerned only with patterns which satisfy criterion 2 in our definition. These may be called the **classical Islamic patterns**, but since there is no possibility of any confusion we shall continue to use the term Islamic patterns to refer to classical Islamic patterns.

1.1.4 Turning to Geometry

The reader will no doubt agree that no art can be appreciated, save at a very superficial level, without some understanding of the concerns and passions of the milieu in which it arose. It is now our purpose to explain the underlying psychological factors which caused Islamic art to seek inspiration in geometry.

It has often been said that geometry was forced on Islamic art by religious proscription. This explanation suffices only at a rather superficial level and misses out that which is deeper and more significant.

Whilst there do exist reported sayings of the Prophet Mohammed, the so called *hadith*, which threaten hell fire to those who would paint or sculpt animate creatures, similar prohibitions also exist in the Bible. "Thou shalt not make any sculpture or likeness of living creatures" — Deuteronomy V, 8., for example. The obvious question may be asked as to why then did such injunctions not stop the followers of the Christian faith from painting likenesses of living creatures?

The most fundamental reason for the substantially non-figurative nature of Islamic art arises from the fact that Islam was born with an all consuming passion to replace anthropomorphic images of God with a single abstraction. For a Muslim, unlike a Christian or a Hindu, God did not and could not become *flesh*. The only material image of God that a Muslim is prepared to employ is that of *Nur*, meaning light. *Allahu Nurus Samavat Bil Arz* — God is the light of the heavens and earth — proclaims the Quran. Since the major concern of art, until fairly recently, has been with images of the perceived deity, and since Islam did not provide any such images, Islamic art embraced geometry.

Driven by the religious passion for abstraction and the related doctrine of unity — *al-twahid*, the Muslim intellectuals recognized in geometry the unifying intermediary between the material and the spiritual world. The works of Euclid and Pythagoras were among the very first to be translated into Arabic. Starting with large scale translations of these in the 8th century, the appreciation of the value of geometry grew and reached high maturity in the following 200 years. We find, for example, the writers of a treatise — *Rasai'il*, who belonged to a 9th–10th century society of scholars known as *Ikhwan Al-Safa* — the Brotherhood of Purity, preaching that[4]:

> ... *the study of sensible geometry leads to skill in all the practical arts, while the study of intelligible geometry leads to skill in intellectual arts because this science is one of the gates through which we move to the knowledge of the essence of the soul, and that is the root of all knowledge.*

We read here a clear statement recognizing geometry as the source which provides for secular as well as religious needs in a unified way.

Thus at the more intellectual levels, the desire for abstraction and the search for unity were two of the main passions which caused the Islamic culture to turn to geometry. There was a third passion.

The most striking characteristic of Islamic geometrical patterns, as we pointed out earlier, is the prominence of symmetrical stars and constel-

---

[4]Quoted by Critchlow [9] from a Translation by S. H. Nasr, see page 7.

lations. Why is there such fascination with star shapes? Is this a mere coincidence? Color Plate 1(a) has been placed at the first position in this book to underline the importance of the answers to these questions to the appreciation of Islamic geometrical art.

The Plate shows the Turkish astronomer Taqi al-Din at work with fellow astronomers in a observatory constructed for him at Istanbul in the 16th century. Although it comes from a period when Islamic astronomy was falling into decline, it nevertheless captures well an important Islamic environment which fuelled the psychology of covering surfaces with star shapes. It also points to the origins of the geometrical skills which were utilized in the construction of the more sophisticated types of Islamic patterns.

The plate shows a variety of astronomical and geometrical instruments and a number of absorbed users. We also see books and individuals engaged in reading and writing. The values and concerns of these characters are evident and reflected in their environment. Since the heavens, which engross them, are entirely covered with patterns and star shapes, so is the local environment created by them.

The attraction of stars is deep rooted in the human psyche. We only have to look at the flags of nations to appreciate how universal is the unconscious wish to *reach for the stars*. All the ancient civilizations — the Babylonians, the Egyptians, the Indians, the Chinese, the Maya, and others assiduously observed the heavens. Some undertook herculean labor in constructing monuments such as the Pyramids to relate to the stars and be guided by them.[5]

The people of the Islamic civilization were not only heir to this ancient tradition, but had even greater need to be guided by the stars. They included many desert populations whose way of life involved extensive nomadic wandering. The Arabs were great sea farers. Whether on land or sea the Muslim had to know, five times a day, the exact direction in which to pray. This was a unique requirement. All this made the stars extraordinarily significant. The Quran abounds with verses which conjure up powerful imageries on the theme — *He it is who hath set for you the stars that ye may guide your course by them amid the darkness of the land and the sea (V:98)*.

The great attraction for star shapes in Islamic art reflects this primordial love and practical involvement with the heavens. From the 9th century

---

[5] It is interesting to reflect that modern physics has concluded that all the elements, including the constituents of living cells, were forged from simple hydrogen through thermonuclear fusion processes within the cores of stars. Perhaps humans have known intuitively, from the earliest of times, that they are in a very true sense *the children of the stars*. Perhaps the continued attraction for Astrology and horoscopes has its origins in this primordial intuition.

onwards, when Ptolemy was translated into Arabic, until well into the 15th century, astronomy was the most passionate intellectual activity in the Islamic world. To quote Nasr [34], see page 98:

> *The works in Arabic and Persian and even other Muslim Languages, such as Turkish, on astronomy and allied subjects are of such great quantity that, despite two centuries of study by Western Scholars, much of the material has remained nearly untouched while a great many works have only been partially analyzed.*

The reader may know that the earliest observatories were constructed[6] and many stars were first named by Muslim astronomers and continue to be known by their Arabic names. Instruments such as the *Astrolabe* and words such as *Almagest* are reminders of the Islamic legacy in astronomy.

Astronomy, the author conjectures, has also been a key source of geometrical skills which were passed on and utilized in the design of complex Islamic patterns. The main basis for this conjecture is the fact that compared to the diagrams that arose in other fields, such as building construction, astronomical diagrams demanded far greater manipulation of complicated structures containing circles. Figure 1.3 shows a typical one. Note the large number of circles of various sizes with lines emanating from their circumferences and the distribution of a set of smaller circles on the circumference of a larger one. These are the kind of structures that underlie the more elaborate Islamic patterns.

To summarize, the passion for abstraction, the search for unity, and the involvement with heavens were the driving psychological mechanisms, which led Islamic art to turn to geometry. It was for these compelling reasons and not simply through a naive fear of a jealous Allah that long before Galileo, the founder of modern scientific method, wrote:

> *Philosophy is written in this grand book, the universe, which stands continually open to our gaze. But the book cannot be understood unless one first learns to comprehend the language and read the letters in which it is composed. It is written in the language of mathematics, and its characters are triangles, circles, and other geometric figures without which it is humanly impossible to understand a word of it; without these one wanders about in a dark labyrinth.*

---

[6] The first astronomical observatory which gathered together eminent astronomers from many lands and conducted extensive observations as well as computations was built at Maraghah in Iran in the 13th century.

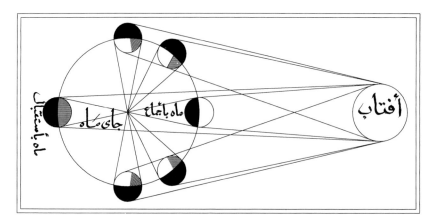

Figure 1.3: An astronomy diagram showing the eclipses of the moon, based on one by the 10th century Muslim astronomer Al-Biruni, see page 106, Nasr [34].

that the Islamic civilization was led to cover its holy as well as secular places with triangles, circles, and other geometric figures.

1.1.5 Related Literature

Finally, to conclude our introduction, we shall give a brief review of the literature. This should be of interest to readers who wish to explore the topic further.

The artists, artisans, architects, geometers, and designers who created and perpetuated Islamic patterns were secretive. They disclosed their methods and discoveries only to a chosen few. The long established tradition where the master reveals his jealously guarded notebooks only to his sons or to a few devoted apprentices is still very much the mode employed today.

Although a set of designs by Mirza Akbar, who worked in the early part of the 19th century as an architect to the Shah of Persia, have survived and are to be found in the Victoria and Albert museum London, and recent researches have unearthed some documents [8] in a few other libraries and museums, no comprehensive treatise on the subject has come down from the more distant past.

The first publication which made available a large collection of Islamic patterns appeared in the later part of the 19th century and resulted from the labors of the Frenchman Prisse d'Avennes [11]. He arrived in Cairo in the footsteps of French scholars who were brought to Egypt in 1798 when Napoleon invaded and initiated a survey of ancient and modern monuments.

Prisse d'Avennes was an extraordinarily talented man. He distinguished himself as a historian, artist, linguist, archaeologist, explorer, and engineer.

1.1. Islamic Patterns: An Introduction

Color lithography had been invented by the German Senefelder in the late 18th century and perfected by the mid 19th. This invention served the needs of Prisse d'Avennes perfectly. Between 1869 and 1877 he published his sumptuous and monumental 4-volume work *L'art arabe* containing 200 color plates which illustrated Islamic designs on mosques, manuscripts, houses, monuments, and other artifacts.

Many nineteenth century European artists, architects, and designers became fascinated by the brilliant colors and abstract geometric designs of Islamic art. One such was the English architect and art decorator Owen Jones who made the first systematic study of Islamic architecture in Spain. He travelled extensively and published another monumental work *The Grammar of Ornament* [21], which surveyed world decorative art. He wrote:

> *The mosques of Cairo are amongst the most beautiful buildings in the world. They are remarkable at the same time for the grandeur and simplicity of their general form, and for the refinement and elegance which the decoration of these forms displays ...*
>
> *When we examine the system of coloring adopted by the Moors, we find that as with form, so with color, they followed certain fixed principles, founded on observations of nature's laws ...*

Most readers would be surprised to learn that it was to record and display the colors of Islamic architecture that color lithography was first employed in book printing in Britain [3]. Owen Jones's treatise on the Palace of Alhambra in Granada, published in stages during 1842–46, was the first color-printed book to be produced in Britain.

Whereas Prisse d'Avennes and Owen Jones published collections of Islamic patterns without too much attempt to offer methods of constructing them, the French architect Bourgoin was the first to try to analyze a large number of patterns for their methods of construction. His pioneering work [6], based on studies of Islamic art in Cairo and Damascus, appeared in 1879.

Relatively recently, several authors — Critchlow [9], El-Said and Parman [13], Wade [50], and Paccard [36] have published large collections of Islamic patterns. Apart from Paccard, the others have offered their own analysis of the methods of constructions. Lalvani [23] is another author who has analyzed a substantial collection of Islamic patterns from India. The insights offered by all of them are interesting and valuable but the reader may find the interpretations of symbolisms offered by some of the authors to be rather subjective. Only Paccard's book, which is very lavish, expensive, in two volumes, and specializes on Moroccan architecture, contains samples of original drawings done by practicing craftsmen and artists. Paccard is also the one who quotes extensively from first hand experience with those who are currently active in the tradition of Islamic pattern design. He is also the one who discusses the zalij approach to Islamic patterns, which is currently the most refined and developed. Lalvani is the only one who has given computer graphics algorithms.

No author so far has been able to offer an account of the evolution of Islamic patterns. How, where, and when did Islamic patterns evolve from the simple to the complex? How did they get transmitted widely, so that the same pattern may be seen in India and Spain? These are fascinating questions to which there are at present no sufficiently detailed answers. Much research is needed.

1.2 GEOMETRICAL STRATEGIES AND STRUCTURES

Our purpose now is to illuminate for the reader the fundamental geometrical strategies and structures behind Islamic patterns. In what follows, we are not intending to repeat what others have done previously, i.e., to offer individual constructions for a large number of patterns. For this, the reader must refer to the literature described in the last section. Our purpose is to distill the essence; to show the reader the wood without forcing him or her to become lost in the trees.

1.2.1 Khatem Sulemani: The Most Basic Shape

Let us start by showing the simplest strategy and the most basic characteristic shape.

The bottom part of Fig. 1.4 shows the most frequently occurring pattern in Islamic culture. It can be spotted as commonly in Samarkand as in the Sahara. It contains the 8-pointed star shape, known as *Khatem Sulemani*, meaning *the Solomon's seal*. This shape, shown prominently in Fig. 1.4d, is the most ubiquitous shape that occurs in Islamic patterns. If we were forced to pick one shape that characterizes Islamic patterns, then it would have to be this one. Not only does it occur with great frequency, but the great bulk of Islamic patterns can be seen as variations on the theme of Khatem Sulemani. This will be explained in the next section. For brevity, we shall refer to this shape as S_8.

Figures 1.4a and 1.4b show how S_8 is based on an 8-fold symmetric division of the circle and the self-superposition of two square polygons.

1.2. Geometrical Strategies and Structures

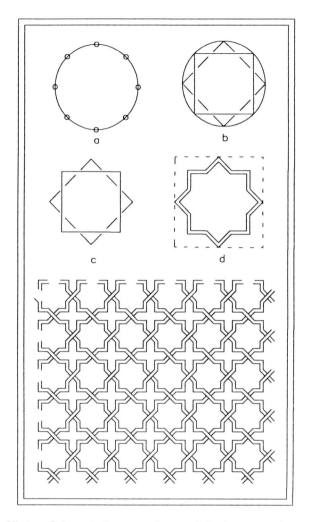

Figure 1.4: Khatem Sulemani, the most characteristic shape, seen here in the most frequently occurring pattern in Islamic culture.

One can, of course, skip the division of the circle and obtain it by just the superpositioning of two squares. This is how it must have been first discovered in antiquity. It demonstrates very simply how **the symmetric division of a circle and the self-superposition of a polygon** give rise to a characteristic star shape, in this case S_8. This basic strategy can be generalized.

The dotted square region in Fig. 1.4d is the **repeat region** of the pattern. It fills space, without leaving holes, when the pattern is drawn. This is what we mean when we say that Islamic patterns of the kind we are discussing in this book are *space filling*.

Note that we can choose to ignore the existence of the square repeat region and construct the pattern simply by tiling, imprinting, or stitching with the shape S_8, as drawn in Fig. 1.4c. These actions would give rise to cross-shaped holes, which, if we employ tiling, could be filled with tiles of that shape.

In yet another approach, we could fill a region of space with contiguous circles, divide the circles as in Fig. 1.4a, obtain the pattern by joining the points as in Fig. 1.4b, and then finally erase the circles.

No matter how we produce the overall pattern, a square repeat region, or equivalently an underlying square grid, is an implicit geometrical property of this pattern.

1.2.2 Variations on Khatem Sulemani

On seeing the two filled patterns in Fig. 1.5, anyone who is familiar with Islamic patterns will instantly recognize them as belonging to that genre. What is shown is the most characteristic visual flavor in Islamic patterns. A large bulk of patterns have this and share a common geometrical structure. We will now demonstrate how patterns with this flavor can be seen as variations on the theme of S_8.

The shapes in Figs. 1.5a and 1.5b have been obtained quite simply by introducing two new concentric circles in Fig. 1.4b. These are again divided symmetrically to yield eight points as was done previously. Joining the points as shown, produces some secondary polygons which have been filled in black. They represent yet other shapes which occur regularly and characterize Islamic patterns.

The radii of the two inner circles are arbitrary and can be varied to alter the size of the inner 8-pointed star shape and the associated polygons. Figures 1.5a and 1.5b show two variations.

Figure 1.5c represents a special case of the shapes that are produced when the radii of the circles are varied. Here, only one of the two interior circles is allowed an arbitrary radius. The radius of the innermost one is forced to have a fixed size in relation to the one in the middle. This is achieved by drawing an octagon through the eight points on the inner circle. The shape is completed by joining lines through the mid-points of the octagon and through making use of the sides of the two squares defined by the outer circle, as shown. The strategy used here reduces the degree of arbitrariness by introducing an inner constraint. The choice of appropriate

1.2. Geometrical Strategies and Structures

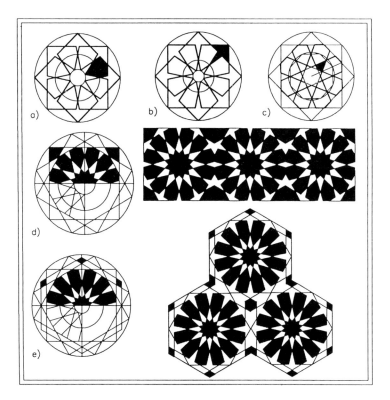

Figure 1.5: Variations on Khatem Sulemani describe the basic geometric structure of the great bulk of Islamic patterns.

inner constraints is the key to achieving harmonious proportions and a sense of beauty. We shall enlarge on this point in the next section.

Note how Fig. 1.5c contains the shape S_8 again at its center. The structure is recursive and we could go on repeating it indefinitely. Note also how the limbs of the star shape, shown in thicker lines, are now defined by lines that are parallel. This was not the case in Figs. 1.5a and 1.5b. This fact, and the fact that the same shape S_8 bounds the interior as well as the exterior of this structure, has made this a favored one in the repertoire of Islamic patterns. We will meet it again when we examine the pattern in Fig. 1.7.

Note also that in any of the cases shown in Figs. 1.5a, 1.5b, and 1.5c we can slice off a square portion to form a repeat cell and use it to fill as much space as we like. The patterns produced in this way will contain 8-pointed star shapes.

It is easy to generalize the above procedure by choosing to inscribe more than two squares in the outer circle. Figure 1.5d uses the same construction as Fig. 1.5a, except that there are now 3 squares, giving 12 symmetric points. We have cut off a square slice from the pattern completed in the circle to form a repeat cell and copied it to produce the filled pattern on the right. The result is a pattern containing 12-pointed star shapes. One could go on in this way by introducing more and more squares.

The pattern shown on the right in Fig. 1.5e is produced similarly to the one above it, except that this time we have cut off a hexagonal repeat region.

1.2.3 Harmonious Proportions: The Secret of Beauty

In the last section, when we examined the diagrams in the top row of Fig. 1.5, we saw that although we can make an unlimited number of shapes with 8-pointed stars, there is one case which is special, the one shown in Fig. 1.5c. This one has the *right* proportions and directions. Its limbs are the cleanest and most pleasing to look at. The construction was achieved by introducing an inner constraint in the form an octagon.

It has been appreciated since antiquity that beauty arises if and only if the constituent parts of a structure are harmoniously proportioned in relation to each other and in relation to the whole. Beautiful patterns have to be based on some form of *inner logic* of proportions. The designers of Islamic patterns have used a variety of strategies to achieve harmonious proportions; some are based on rules of thumb and have been discovered through experience. However, there is one systematic strategy which has been used extensively. This involves the structuring of space with a series of nested polygons. Figure 1.6 shows an example.

Referring to the top left diagram in the figure, it is easy to see that $AD/ad = 2/\sqrt{2} = \sqrt{2}/1$. Thus, relative to the outer square, the side of the inner square is reduced by the factor $1/\sqrt{2}$ and the area is halved. This nesting procedure can be executed repeatedly to obtain a whole series of squares as shown in the middle diagram. The scheme is often referred to as the **Root Two System of Proportions**; see for example page 8, El-Said and Parman [13].

An allied scheme, shown on the top right, ignores the rotated square members and selects only the ones with the same orientation as the outer most one. In this case, the side of each inner square is halved relative to the outer and the area is reduced by a factor of 4.

A typical design which utilizes the first scheme is shown in the bottom diagram in Fig. 1.6. Note how, inside the repeat cell, the lines elements in the motif are forced to coincide with lines forming the sides of the nested

1.2. Geometrical Strategies and Structures

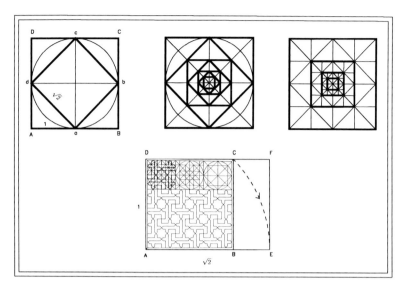

Figure 1.6: The Root Two System of proportions.

squares. The beauty of the pattern comes from the satisfaction of these constraints and the underlying logic of proportions.

The proportion $1 : \sqrt{2}$ is also enforced globally to obtain overall beauty. Typically, the dimensions of a rectangular wall or a floor are chosen so that the sides of the rectangle conform to this ratio, as is true in the case of the rectangle ADFE in Fig. 1.6. In a tiled wall, a sense of beauty is achieved by covering the square region ADCB with a pattern whose internal structure conforms to this ratio, the rectangle BCFE being utilized for calligraphy and border elements.

Ratios, such as $\sqrt{3}/1$, the golden ratio $(1 + \sqrt{5})/2$, and others arise when other polygons are used in nesting schemes similar to the one shown in Fig. 1.6. The interested reader is referred to El-Said and Parman [13] and Critchlow [9] for examples of these. Paccard [36] and Wade [50] contain many examples of rules of thumb schemes which are utilized to achieve harmonious proportions.

1.2.4 A Complex Pattern

We now move on to Fig. 1.7 which reveals the structure of a more complex pattern than we have seen so far. The pattern is the one shown in the color Plate 3c. It contains as sub-shapes a characteristic shape we met earlier. The inner logic of this pattern is based on the packing of space with

Figure 1.7: A complex and beautiful pattern.

tessellating polygons.

This pattern relies on 16-fold divisions of circles placed at the center and at the vertices of a square repeat cell. The repeat cell itself has 8-fold symmetry and, therefore, it suffices to show the construction lines in one eighth of the square. This we have done on the bottom left. The construction procedure as illustrated is based on one published by Hankin [16].

The bottom region on the right shows the first move which produces a well-positioned octagon. As shown on the left, the octagon is then used to construct five pentagons and two half hexagons. This structure is the basis of the inner logic of the pattern. The polygons are drawn by extending the radii of the octagon and terminating them as shown. The pattern is created by drawing lines which join the mid points of these polygons. Note

1.2. Geometrical Strategies and Structures

how the shape S_8 arises again through the use of the underlying octagon, exactly as it did in Fig. 1.5c. The shape that appeared in Fig. 1.5c is a sub-shape of the repeat cell of this pattern.

1.2.5 Grids and Circles

So far, in all our examples, we have started with the construction of a repeat cell. The grids which have arisen when the cell has been copied have been implied, but have played no significant role. Although in theory, the two approaches are equivalent, some patterns can be constructed more easily by starting with a grid and allowing the repeat cell to appear implicitly.

The grids used most commonly are the isometric grid (made from equilateral triangles), the square grid, and the rectangular one. Since six equilateral triangles combine to form a regular hexagon, the isometric grid can also be considered as being hexagonal. Similarly, since the rectangular grid can be divided into isosceles triangles or rhombuses it may alternatively be thought of as being triangular or rhombic, as shown in Fig. 1.8. Complex patterns sometimes employ more complicated grids, as we shall see.

We now show two examples where the grids are made more explicit. The pattern in Fig. 1.8 is based on a rectangular grid. The strategy for its construction is as follows:

Start with a circle (consider the first complete circle at the top) and divide its circumference into 10 equal arcs. Now, draw two diameters to pass through points on the circumference as shown. Extend the diameters equally in the four directions, joining the end points to form a rectangle.

Next, copy the original circle at the corners of the rectangle. The rectangle and the circles placed at the corners and at the mid-point of the rectangle can now be copied repeatedly. This results in a rectangular grid whose nodes are occupied by marked circles.

The rest of the construction involves joining the points on the circle symmetrically as illustrated in the figure. The repeat region of this pattern, which was not chosen as the starting point, is a rhombus obtained by joining the centers of four adjacent circles. One such cell is shown in bold. Note that this pattern contains 10-pointed star shapes, which arise mostly on a rectangular grid.

Figure 1.9 shows a pattern based on a grid which combines squares and equilateral triangles. This grid is, in fact, one of the eight referred to by mathematicians as a *semi-regular* or an *Archimedean* tessellation. The construction method packs circles densely and employs division into 7. Steps in the construction of the pattern are shown.

22   Chapter 1. Islamic Patterns and Their Geometrical Structures

Figure 1.8: Diagram showing the construction of a 10-pointed star pattern on a rectangular grid.

1.2. Geometrical Strategies and Structures

Figure 1.9: A pattern made from a dense packing of space with circles, on a grid made from squares and equilateral triangles. Such packing arises in atomic and molecular structures in the real world and makes Islamic patterns of interest to physicists, chemists, and crystallographers.

Note that the diagram in Fig. 1.9 resembles the kind that we meet in physics when we study phenomena at the atomic level. The symmetric divisions and the dense packing of space with circles is a feature of Islamic patterns which relates them to the kind of structures that arise in the physics of solids. We shall return to this theme in the next chapter.

1.2.6 The Zalij Approach to Islamic Patterns

In our discussion above, we have already pointed out several polygons as representing characteristic shapes. Such shapes arise frequently and produce the Islamic flavor in patterns.

In Morocco, where the traditional art of Islamic patterns is to be found in its most refined and developed form, the approach to pattern making is almost entirely based on the utilization of sets of characteristic shapes and sub-shapes as mosaics. Long experience with tiling has led to the recognition and evolution of such shapes and their relative proportions. The shapes are hand cut by craftsman from enameled terracotta squares and are known as zalij. Some typical zalij shapes are shown in Fig. 1.10.

Zalij panels are constructed in brilliant colors and can be seen all over Morocco in public as well as private places. The art of zalij enjoys the

Figure 1.10: Some zalij shapes.

1.2. Geometrical Strategies and Structures

highest patronage. His Majesty King Hassan II, himself, has given strong personal support to promote, preserve, and develop the art. The city of Fes is the most prominent center and the spiritual home of the art, but it is widely appreciated throughout Morocco. A fine example of zalij tiling on a fountain outside the mausoleum of Mohammed V in Rabat is shown in Plate 10. The interested reader should refer to the book by Paccard [36] to enjoy a feast of zalij. Another sumptuously illustrated book on the subject is the one listed as reference [53].

The designers of zalij patterns, known as *Mallems* create their designs on graph paper using grids which have evolved from experience. They pack space with zalij shapes in the manner of a jigsaw puzzle. Their method will now be illustrated.

Consider the design on the graph panel in Fig. 1.11. This was done by the author to produce a repeat cell containing a star shape with 24 rays

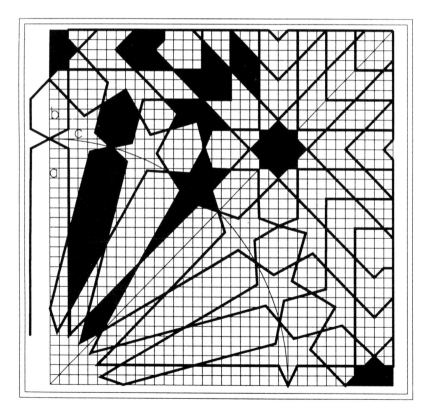

Figure 1.11: Diagram showing the initial construction of the zalij design completed in Fig. 1.12.

as shown in Fig. 1.12. The design on the graph uses a 36 × 36 grid. This choice is made to allow for divisions into 12 × 12 sub-squares.

The first element in the design is the circle which passes through the vertex of the polygon marked **a**. This circle has a radius of 25 divisions. It defines the polygon **a** which is superimposed on itself to produce the large star shapes at the center. The logic of the choice 25 for the radius is that this makes the straight edge of the polygon 24 divisions long. An extra division is used to close the polygon to give it a zalij shape.

Polygon **b**, which is another zalij shape, is constructed by drawing lines as shown. The rotations of the polygons **a** and **b** and extensions to the sides of **b** produce the zalij marked **c**. The rest of the design is completed by drawing standard zalij shapes on the grid.

For example, the shape S_8 was constructed with its center at the intersection point of the vertical and horizontal tangents to the circle which

Figure 1.12: A typical zalij panel.

happens to be a corner of one 12 × 12 sub-square. It spans 6 divisions on the grid. The shape was produced by drawing lines which run one division parallel to the grid, followed by one division at 45 degrees.

All other shapes were completed in this way by drawing lines which run either along the grid lines or at 45 degrees to them. Some of the characteristic zalij shapes have been filled in black.

1.3 CONCLUDING REMARKS

In this chapter we have introduced the reader to Islamic patterns and their recognizable characteristics. We have explored the intellectual, psychological and practical reasons which led to their creation. We have also tried to give insight into the geometrical strategies and structures that underlie Islamic patterns. Finally, we have pointed out a number of characteristic shapes that occur in Islamic patterns and shown how they are currently used in the art of zalij.

We have noted that the most basic geometric features of Islamic patterns arise from symmetric divisions of circles and replication of unit cells. We shall return to the significance of these simple features in the next chapter.

Chapter 2

IN PRAISE OF PATTERN, SYMMETRY, UNITY & ISLAMIC ART

The geometric spirit is not so bound up with geometry that it cannot be disentangled and carried into other fields. A work of morals, of politics, of criticism, perhaps even eloquence, will be finer, other things being equal, if it is written by the hands of a geometer.

Fontenelle

The last chapter introduced the reader to Islamic patterns and the geometrical structures behind their symmetries. This chapter will begin by trying to convey something of the immensity of the notions of pattern and symmetry in general terms. We shall then attempt to show how the most creative amongst scientists are driven by a hunger for beauty and art. This will establish for us the relevance of the Islamic tradition which offers the gifts of pattern symmetry and unity.

In the last part of this chapter, we shall point out how Islamic geometrical patterns are particularly suited to the celebration of symmetries of the kind that occur at molecular levels of reality. We shall create a new work of 'science-art' to honor this unity.

2.1 IN PRAISE OF PATTERN

> *Life makes patterns out of patternless disorder, but I suggest that life itself was made by a pattern and that this design is inherent in cosmic forces to which life was, and still is, exposed.*
>
> Lyall Watson

It is not possible for the human brain to imagine a state comprising only a jumble of random and disordered events. Over aeons of time, out of chaos and disorder, life and consciousness have emerged in the universe — together with pattern and order. Pattern is everywhere.

There is pattern in the sky. There is pattern in the sea. There are patterns in sand dunes. There are patterns in thought, language, poetry, music, social behavior, peacock feathers, the singing of the birds, the dancing of the dervishes, the binomial theorem and the equations of mathematical physics — to mention just a few instances.

The notion of pattern is central to understanding. Above all, the brain is a recognizer of patterns. Ultimately, every art and every science is based on patterns, if we define pattern to mean *any regularity that can be recognized by the brain*.

The extraordinary progress we have made in the last few hundred years in understanding the laws of nature has come through our increased understanding of patterns. In his classic book *The Mysterious Universe* [20], Sir James Jeans wrote,

> *Our remote ancestors tried to interpret nature in terms of anthropomorphic concepts and failed. The efforts of our near ancestors to interpret nature on engineering lines proved equally inadequate ... On the other hand, our efforts to interpret nature in terms of the concepts of pure mathematics have, so far, proved brilliantly successful. It would now seem to be beyond dispute that in some way nature is more closely allied to concepts of pure mathematics than to those of biology or of engineering.*

Why is this so? This is so because pure mathematics, which most people erroneously identify with numerical calculations, is in truth concerned with classification and study of patterns. As the pure mathematician G. H. Hardy asserted [17]:

> *A mathematician like a painter or a poet is a maker of patterns.*

2.1. In Praise of Pattern

Another mathematician, Herbert Turnbull, made the same point in his book *The Great Mathematicians* [49] when he said this about Pascal:

Numbers and quantities are not so important for their size or bulk as for their patterns and arrangements. What Pascal did was to bring the notion of pattern, common enough in geometry, to bear upon number itself — a highly significant step in the history of mathematics. By so doing he created higher algebra and prepared the way for Bernoulli, Euler and Cayley.

Highly significant steps resulting from a shifting of focus onto patterns and arrangements have occurred and continue to occur in the history of every science and not only in the history of mathematics.

Chemistry experienced its greatest revolution when the Russian chemist Dmitri Mendeleyev put the elements into a 'pattern' which became known as the periodic table. From the position of an element in this pattern it then became possible, for the first time, to predict the chemical behavior and the properties of elements. Mendelev, from his pattern alone, was able to predict the properties of ten 'new' elements which at that time had not been observed or suspected of existence. Today, it has been said that a chemist can achieve as much without the periodic table as a sailor can without a compass.

More recently, the discovery of the DNA 'pattern' has revolutionized molecular biology and revealed the most astonishing truths about life. All that makes a human a human, an insect an insect, and an elephant an elephant is a consequence of the pattern stored as the genetic code. The power placed in the hands of humans through acquiring an ever increasing knowledge of this pattern is awesome. The future consequences of being able to modify this pattern cannot be grasped.

Apart from mathematics, only computers enjoy a universality of application. There is only one tool we have invented so far which is so versatile that it can fly spaceships, design artificial kidneys, zap nine monsters, check the football results, and also predict the hole in the ozone layer.

What makes the computer so infinitely adaptable? The answer, as in the case of mathematics, has to do with patterns. A computer is of universal application only because it first reduces all information, be it to do with spaceships or artificial kidneys or monsters or football results or the hole in the ozone layer, into patterns of binary digits. It then proceeds to process these patterns with accuracy and speed.

Of all the things in the universe we seek to understand, the workings of the human brain are the most complex and challenging. Modern research into neural networks has pattern recognition as its key quest. From success in constructing pattern-recognizing computers we will be able to design even more versatile machines than those of today. Some of these will be used to seek out, explore, and understand yet more complex and graceful patterns which remain hidden in every nook, cranny, and layer of the universe. The world of tomorrow will continue to be transformed as we grow more and more in our ability to create, manipulate, and recognize patterns.

Such then is the immensity and significance of the notion of pattern. Without patterns the whole edifice of consciousness would fall to pieces. There will be void.

2.1.1 In Praise of Symmetry

Symmetry deserves to be the subject of periodic festivals, so that in its lofty atmosphere one can search for the relation between different manifestations of the human spirit.

Kh. S. Mamedov

Symmetry, like pattern, is omnipresent. It is the glue which binds the Universe. It binds it physically, aesthetically, morally, and in all kinds of other ways — some obvious, some remaining mysterious.

The force of gravity which determines the large scale structure of the Universe is radially symmetric and so is the electrical force which holds electrons in orbits around the nucleus. Symmetry is seen in the rings of Saturn and in the skins of soap bubbles. It occurs in gigantic swirling tornados and in the splashing of microscopic milk drops.

The concept of symmetry, like that of pattern, is an extremely wide ranging one. Its significance in the context of spatial structures and visually striking patterns is easily appreciated. But as we shall see in the next chapter, symmetry has incredibly wide ramifications and domain.

In its every day usage the term symmetry is most often employed to describe balance or the exact correspondence of size and shape between opposite sides of a structure. More generally, it may refer to other kinds of regularities which are displayed by objects which are made up of identical or similar parts. The term is also used to refer to harmony of proportions.

In science, symmetry, unlike pattern, has a precise meaning. This we shall meet in the next chapter. Here, we shall use the term without bothering to define it and assume that the reader will not be at a loss.

2.1. In Praise of Pattern

Most of all, symmetry is associated with that which is beautiful, eye catching, and perfect.[1] Symmetry has cast its spell on the brain from the earliest of times. All living organisms — flowers, plants, insects, fish, birds, and others — use symmetric patterns, sometimes quite stunning ones, to attract.

Experiments have shown that female birds and zebras prefer males with more symmetric markings. Human ornaments and artifacts from every culture and every age abound in symmetric patterns. The evidence for the human fascination with the symmetry of the form of the opposite sex is self evident.

In literature and philosophy, symmetry has often been used as a metaphor to express other kinds of attractions. Aristotle, for example, compares a good and virtuous man to a perfect cube. The Chinese have used the symmetric Yin-Yang symbol to express their philosophic notion of harmony.

The Pythagoreans considered the square and the circle to be possessed of special powers because of their perfect symmetry. Ptolemy could only conceive the planetary orbits as being circles and Kepler spent much labor in trying to establish a connection between the planetary orbits and the symmetric platonic solids. Such views, naive though some of them turned out to be, have arisen from intuitive grasps of some very profound truths.

Why is symmetry so eye catching? There has to be some physiological explanations. Experiments by psychologists have shown that the eye-brain system is specially tuned for the detection of symmetry. The presence of symmetry is grasped far more rapidly and accurately than its absence and causes physiological arousal which we interpret as pleasurable feelings. But why is this so? The brain must have evolved in this way in response to some basic needs of survival. Some of the reasons for the attractions of symmetry are not difficult to fathom out.

To survive, the brain has to be able to recognize security and stability. The existence of symmetry is clearly associated with these. An asymmetric three legged elephant will not manage to endure in the forest for long. Indeed, one cannot imagine any truly resilient structure which does not possess a high degree of symmetry. Since survival is of paramount concern, and since objects which endure have symmetry, one should not be surprised to learn that the brain is tuned to respond to symmetry.

In living organisms, symmetry is connected with health in mysterious ways. Symmetric flowers produce more nectar than asymmetric ones. In

---

[1] We need to remind ourselves that an excess of symmetry is not necessarily beautiful — think of a centipede. Similarly, a perfectly symmetric circle in not as interesting as an 8-pointed star, say. Ian Stewart and Martin Golubitsky in their book *Fearful symmetry* [46] have emphasized the notion of the *breaking* of perfect symmetry.

humans, various kinds of asymmetries reflect the likelihood of different diseases. For example, asymmetries in fingerprints can be used reliably to predict the possibility of a person succumbing to schizophrenia. Why is fivefold symmetry so preferred in living forms?

Another reason for the existence of the 'symmetry response' in the brain must be associated with the need to grasp and retain information. Just as rhyme and poetry are easier to remember compared to ordinary text, so patterns that have a high degree of symmetry are easier to remember than those that lack them. They contain redundant information and can be stored more economically. Symmetry offers efficiency in grasping, storing and recalling information.

Symmetry gives the brain the power to predict. This is another important quest in survival. If we can see only a small amount of a symmetric structure then we can predict the rest. The greater the symmetry the smaller the amount of information needed to make predictions. The power to predict gives a sense of confidence and security. This power has been used to great effect by modern physicists.

The fact that nature prefers symmetry has been exploited by theoretical physicists to make many powerful predictions and develop deeper theories. For example, James Clerk Maxwell introduced a term in his electromagnetic equations purely on symmetry grounds which led to the prediction and understanding of electromagnetic waves. In the early part of this century Paul Dirac predicted the existence of antiparticles on symmetry grounds, a prediction which was first treated with great skepticism. Very soon, however, the positron, i.e., the anti-electron, was observed and twenty years later came the discovery of the antiproton.

Physicists have now discovered that all fundamental particles of matter have a mirror twin with the same mass but opposite charge. Chemists have discovered that many organic compounds are chiral, that is, they exist as two mirror image species termed enantiomers. All the recent triumphs of theoretical physics have relied on considerations of symmetry.

Symmetry also offers economy in manufacture. A symmetric object is made up using the same component repeatedly. Nature, it is clear, prefers economy of construction utilizing the same components over and over again in different combinations.

A recent interesting discovery in the subject is that the symmetry response of an individual is not entirely genetic but seems to be strongly influenced by his environment. Dorothy Washburn, an archaeologist at the University of Rochester, and Donald Crowe, a mathematician at the University of Wisconsin in the USA, have carried out extensive studies of symmetry in patterns and designs produced in various cultures. Their recent book *Symmetries of Culture* [51], which has received much publicity, announced their remarkable finding that the choice of arrangements of motifs in cultures to produce symmetric designs is not at all random. They discovered that only certain symmetry types are preferred and intuitively recognized as being right by each culture — *"the designs in any given culture are organized by just a few symmetries rather than all classes of the plane pattern symmetries."*

The discovery of Washburn and Crowe adds new significance to the study of the symmetry type of each culture. Symmetry analysis of designs, which will be discussed in chapter 4, is now being adopted by many archaeologists and art historians as a more objective method of analyzing cultural style than methods that have been used in the past. We shall present our discovery of the preferred symmetry type in Islamic cultures in chapter 5.

2.2 IN PRAISE OF UNITY

So powerful is the light of unity that it can illuminate the whole Earth.

Baha Ullah

A story which has often been related in scientific circles concerns the British Scientist Paul Dirac, who in the early part of this century played a key role in developing the theoretical foundations of quantum mechanics. In 1956, the great man visited Moscow University where after a seminar he was implored to write down just one single message on a blackboard for posterity. Without hesitation, Dirac picked up the chalk and inscribed:

A physical law must possess mathematical beauty.

If there were any in the audience who had read their Keats, they would undoubtedly have been reminded of the line:

Beauty is truth, Truth is Beauty.

The Dirac story captures something which remains almost entirely unappreciated, not only by the public at large but also by those who set out to educate future scientists. Contrary to popular belief which holds science to be cold and austere, the moving power of science resides in the imagination rather than in reason. At its most creative, the lungs of science thrive on the same oxygen as those of art — the oxygen of beauty and mystery.

Although our culture perpetuates a cold mechanical image of the scientist, the evidence shows that the most creative rational thinkers do not live by logic and experimentation alone. Many of the most eminent have attested to this.

Typically, Hermann Weyl, an illustrious mathematician wrote:

> My work has always tried to unite the True with the Beautiful and when I had to choose one or the other, I usually chose the Beautiful.

Einstein remarked:

> The most beautiful experience we can have is the mysterious. It is the fundamental emotion which stands at the cradle of true art and true science.

Evidently, the hunger pains for beauty and truth are felt simultaneously. It follows that if we are to nurture creative scientists, then offerings of food for thought alone will be not enough. We also have to offer ample nourishment for the imagination. We must offer art and science in unity.

2.3 IN PRAISE OF ISLAMIC ART

Islamic art is the supreme triumph of pattern ...

David Talbot Rice

Islamic art, as we saw in the last chapter, focuses on pattern and symmetry. Through them it offers unity in science and art. Our object in this chapter, so far, has been to show that these are invaluable gifts. True, the notions of patterns and symmetry explored above were not intended to be confined to visual geometric objects. But more general or abstract notions of pattern and symmetry can best be introduced through visual geometrical ones. As Leonardo da Vinci pointed out:

> The eye, which is called the window of the soul, is the chief means whereby the understanding may most fully and abundantly appreciate the infinite works of nature.

2.3. In Praise of Islamic Art

In its traditional form Islamic patterns are limited to Euclidean geometry but this is a plus. It is in the context of Euclidean geometry that the visual experience is the most important one in early education. Euclidean geometry was the first intellectual discipline to develop a truly mature logical structure. It remains the best vehicle for introduction to logical thought. Furthermore, geometry is valuable for much more than simply learning about logic and proving theorems. It has much broader relevance in education. Robert Dixon gave the reasons [12]:

> *Geometry has more value than simply as a means of acquiring a grasp of mathematical concepts. The applications of a disciplined spatial intuition to art and design, and to the study of morphologies in every conceivable science is so great that perhaps we might think of geometry as a semi-autonomous department of mathematics with different as well as overlapping purposes to abstract mathematics.*

The Islamic tradition in art is, for the reasons given, a worthy and relevant one. It deserves to be cherished and developed. The treasury of Islamic geometrical patterns can be used to serve the needs of education and those of cultured, creative, civilized human beings everywhere.

We shall explore how geometrical patterns provide a natural path to some powerful abstract ones in the next chapter. This chapter will conclude by examining an area where Islamic geometric patterns are particularly suited to offering a unified experience of science and art.

2.3.1 Islamic Patterns and Atomic and Molecular Structures

In the last chapter we showed how the geometric structures of the bulk of Islamic patterns are based on the symmetric division of concentric circles placed on nets, the tight packing of space with polygons and replication. These simple processes are the two-dimensional analogues of some of the most basic processes of nature at the atomic and molecular levels. Symmetry diagrams similar to the ones that arise in the construction of Islamic patterns arise in physics and chemistry in particular, but also in biology.

Examine Fig. 2.1. It shows some typical diagrams used by physicists and chemists to describe atomic and molecular phenomena. Diagram (a) arises in a description of the energy levels of the hydrogen atom. Diagram (b) shows a germanium atom. Diagram (c) shows a benzene molecule and diagram (d) shows the structure of C_{60}, a recently discovered form of

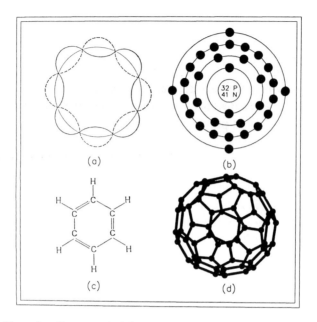

Figure 2.1: Examples of some typical diagrams used by physicists and chemist in representing atomic and molecular phenomena.

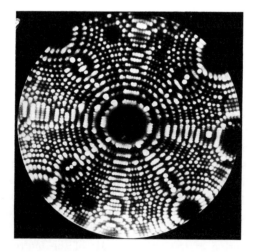

Photo 2.1: A view of atomic structure in a crystal lattice, obtained using a field ion microscope (courtesy Dr. Chen and Professor Messier, Penn State University, U.S.A).

2.3. In Praise of Islamic Art

a carbon molecule named Buckministerfullerene. Here, we have the same basic geometrical structures which we met in chapter 1.

Replication, which is a feature of Islamic patterns is also a feature of extraordinary significance in nature at microscopic levels. The atomic arrangements in solids are very similar to a mosaic. Atoms or clusters of atoms appear in repeating motifs in exactly analogous manner to the tiles in a mosaic. Photo 2.1 shows an image of an atomic structure in a crystal. Note the symmetrical distribution of circles.

Replication is almost synonymous with life. How did life begin on Earth? We cannot give an exact answer so far. But what we can say is that somehow, about four thousand million years ago, a molecule learnt to replicate itself. The simple act of replication is the most miraculous in the whole Universe.

Not surprisingly, for the reasons just given, several authors, amongst them Critchlow [9], Mamedov [30] and Makovicky [26], have noted that the molecular structures of several naturally occurring substances, e.g., $Be_3Al_2Si_6O_{18}$, known as Beryl, are identical to that of an Islamic pattern (see Page 90 Nasr [34], for example). For the same reasons Islamic patterns, produced hundreds of years ago, continue to be of great interest to crytallographers; see for example Makovicky [29]. They are particularly suited to discovering art and science at molecular levels of reality. Here is an example.

2.3.2 A Celebration of Unity

Examine Fig. 2.2a. This shows the structure of the molecule $[Fe(OCH_3)_2 \times (O_2CCH_2Cl)]_{10}$, known more simply as the Ferric Wheel. The molecule was synthesized by the American chemists Stephen J. Lippard and Kingsley L. Taft in 1990 in their efforts to understand certain chemical reactions that occur in biological systems. Note the symmetric tenfold division of the circle and the radial distribution of hexagons to produce a 10-pointed star shape.

Figure 2.2b is a diagrammatic representation of one of the strands of a DNA double helix, viewed down the helix axis. Note again the tenfold symmetry and the radial distribution of hexagons to produce a 10-pointed star shape.

Structures of the kind shown in Fig. 2.2 are the type that arise in Islamic patterns. They are strikingly beautiful.

Writing in the *Scientific American* of February 1993, the chemist Roald Hoffmann, who won the Nobel Prize in Chemistry in 1981, had this to say about the molecule Ferric Wheel [19].

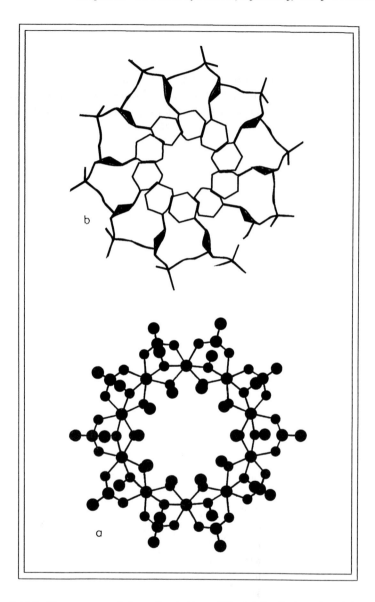

Figure 2.2(a): The structure of the molecule Ferric Wheel (after K. L. Taft and S. J. Lippard [48]). (b): The structure of one of the strands of a DNA double helix, viewed down the helix axis (after L. Stryrer [47]).

2.3. In Praise of Islamic Art

> — *for me, this molecule provides a spiritual high akin to hearing a Haydn piano trio I like. Why is this molecule beautiful? Because its symmetry reaches directly into the soul. It plays a note on a Platonic ideal. Perhaps I should have compared it to Judy Collins singing "Amazing Grace" rather than Haydn trio. The melodic lines of the trio indeed sing, but the piece works its effect through counterpoint, the tools of complexity. The Ferric Wheel is pure melody.*

The author has noted that the structure of Ferric Wheel is very similar to a 'molecule' in an Islamic tiling described by Critchlow [9] (see page 88). This tiling is shown in Fig. 2.3. Compare the black filled region with the Ferric Wheel structure in Fig. 2.2.

Inspired by the words of Professor Hoffmann, the author has constructed several new patterns based on the structure of the Ferric Wheel. One example in shown in Fig. 2.4. Another one, rendered more artistically, is shown in Fig. 2.5 and has been named **The Islamic Ferric Wheel**. The same has been rendered in color in plate 11. Three dimensional designs based on the Islamic Ferric Wheel and done in collaboration with Greg Edwards of Silicon Graphics are shown in color plate 12.

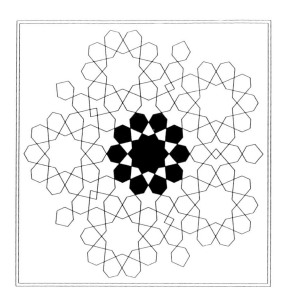

Figure 2.3: An Islamic tiling (after Critchlow [9]).

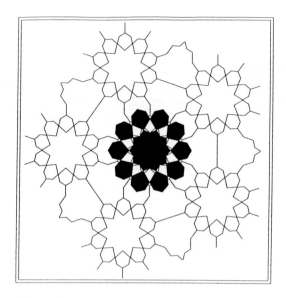

Figure 2.4: A new Islamic pentagonal pattern inspired by the Ferric Wheel.

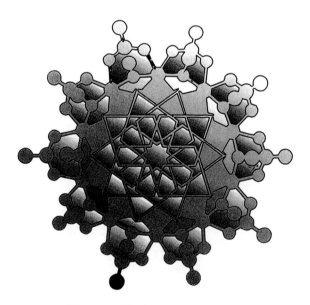

Figure 2.5: The Islamic Ferric Wheel.

2.4 CONCLUDING REMARKS

In this chapter we first explored, in very general terms, the immense significance and range of the notions of pattern and symmetry. In visual geometrical form, these notions span science and art and allow for experiencing the two in unity. We pointed out how science at its most creative is driven through the search for such experience and how Islamic art is particularly suited to offer such experience in explorations of molecular levels of reality.

If art and science are to be practiced in unity, beyond a superficial level, then the visual experience of beauty and mystery must lead to abstraction and to symbolic mode of expression. It should lead to precision and enunciation of rational truths. In the next chapter we will discuss how Islamic art may be used to provide a lead into some of the most powerful abstractions and symbolisms of present day mathematics.

Chapter 3

THE GATEWAY FROM ISLAMIC PATTERNS TO INVARIANCE AND GROUPS

> *The theory of Groups is a supreme example of the art of mathematical abstraction.... Group theory has also helped physicists penetrate to the basic structure of the phenomenal world, to catch glimpses of innermost pattern and relationship.*
>
> J. R. Newman

Geometry, like any other branch of mathematics, gives much more than is asked of it. When the Greek geometer Menaechmus sliced a cone to study the properties of the curves which later became known as conic sections, he had no way of foreseeing that more than two thousand years later these same curves would be needed for understanding the paths of the planets. Similarly, when Muslim artists set out to explore two-dimensional symmetric periodic patterns, they could not have foreseen that several hundred years later symmetry would turn out to be central in an awe-inspiring range of human intellectual endeavor.

In the last chapter, we have already sung the praises of symmetry. We also showed an example of how Islamic art was particularly suited to offering a unified experience of science through the discovery of symmetry at molecular levels of reality. But the discussion in the last chapter was restricted to the visual experience of symmetry. Our purpose in this chapter is to show a glimpse of the immensity of the abstract notion of symmetry and the even greater unity that it has to offer.

In this chapter we will develop this abstract notion of symmetry. We will show, without making any demands on mathematics beyond school level, how Islamic geometrical art provides an ideal gateway which leads directly and naturally to group theory, the mathematical theory most often used to describe symmetry. We will explain the principles which give power to group theory and to the abstract notion of symmetry. The reader should obtain a broad understanding of what has led symmetry to provide the foundation stones on which some of the loftiest citadels of modern science rest.

The chapter will also give a broad outline of the method used by us in our computer graphic studies of Islamic geometrical patterns. This will prepare the reader for the next chapter where more details will be given.

We shall begin our task by examining the symmetries of a very simple and commonly found Islamic pattern. This will be used to illustrate how patterns may be analyzed, generated, and appreciated through their symmetry properties. This will also provide us with an opportunity to point out two of the most central features which characterize science.

Compared to the methods of elementary geometry, which were employed originally in the design of Islamic patterns, this approach is based on a deeper recognition of key spatial relationships that exist in geometrical patterns. It is these relationships which provide a link with one of the most powerful abstractions that the human mind has so far produced. This link will be explored in the rest of the chapter.

3.1 AN EXAMPLE OF SYMMETRIES OF AN ISLAMIC PATTERN

Examine the top part of Fig. 3.1. This shows a pattern found very commonly in Islamic designs. It occurs on the first page of Bourgoin's collection [6]. It can be made very easily by joining vertices on a hexagonal grid.

Let us now explain how this pattern can be analyzed and constructed through the recognition of its symmetries. The pattern here is of a type known as p6m and belongs to the most popular class of patterns found in Islamic art — see Fig. 5.1. The meaning of the symbolic notation p6m will

3.1. *An Example of Symmetries of an Islamic Pattern* 47

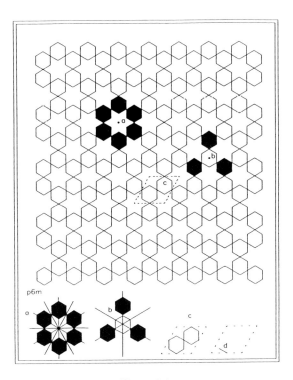

Figure 3.1:

be explained in the next chapter, but for the time being we will just state the key identifying features of a p6m type pattern.

Any p6m type pattern has the property that it possesses 6-fold centres of rotation. By this we mean that if we traverse an arc of 60 (=360/6) degrees around such a center then we arrive at a position which is geometrically identical to the starting position. The center of every star shape in the pattern is a 6-fold center of rotation. This is easily appreciated, for example, by examining the region around the star shape marked **a** in the pattern. A small portion surrounding this particular star has been filled to bring out the 6-fold rotational symmetry around it. It should be appreciated that the center of every star has this property.

Another property of a p6m pattern is that through every center of six fold rotation there are to be found 6 sets of mirror reflection lines. To avoid cluttering the pattern, the sub-region marked **a** has been extracted and redrawn at the bottom left where all the mirror lines through the center are shown. Again, it should be appreciated that such sets of mirror lines pass through every 6-fold center of rotation in the pattern. Although a p6m pattern has some other centres of rotation and lines of reflection, e.g., 3-fold centres of rotation with 3 sets of mirror lines — such as the one marked **b**, the discovery of 6-fold centres of rotations with 6 sets of mirror lines passing through them is sufficient information to identify a repeating pattern as being of type p6m. Repeating patterns, as the name suggests, are those which can be constructed by repeatedly copying some basic motif on a regular grid, i.e., a grid defined by points of intersection of two sets of parallel lines.

Various portions of a given pattern can be chosen to construct the complete pattern by simply copying that portion repeatedly. However, we can always find a parallelogram-shaped region which has the minimum possible area and which is, therefore, the most economical one that may be chosen. For some types of symmetric patterns such a parallelogram may in fact be a special one — a rectangle, a square, a rhombus or a special rhombus whose sides include an angle of 60 degrees. Such a parallelogram region is called a **unit cell** or a **lattice unit** or a **repeat unit** of that pattern.

The dotted parallelogram region marked **c** is a unit cell of the pattern in Fig. 3.1. In this case the parallelogram cell is a rhombus whose sides include an angle of 60 degrees. All p6m type patterns have this special rhombus as their unit cell. We shall refer to the geometrical motif contained in a unit cell as a **unit motif**. The unit motif of this pattern is the double hexagon shape drawn in solid lines in the diagram marked **c** at the bottom of the pattern.

Because of the fact that a unit cell contains various centres of rotation and lines of reflection, the unit motif it contains can be constructed by performing reflections and rotations on a part of itself. This part, we shall refer to as the **template motif**. This can be thought of as the heart of a repeat pattern. For the pattern in Fig. 3.1, it comprises just the little solid line marked **d** in the bottom diagram. The information contained in the infinite pattern of Fig. 3.1 has now been reduced to the minimum possible in the line **d**. Given this line **d**, i.e., the template motif and the knowledge that it is a p6m pattern, we can create as large an extent of the pattern as we like. Let us see how.

Examine Fig. 3.2a. This shows the positions of mirror reflection lines and the centres of 6-fold and 3-fold rotations in a unit cell of a p6m type pattern. The double lines are the mirror lines, the dark triangles and hexagons show centres of 3-fold and 6-fold rotations, respectively. Note how the mirror lines divide the unit cell into 12 triangular sub-cells. For a p6m type pattern any of these sub-cells can serve as a **generator region**, also called a **fundamental region**, i.e., a region which contains a template motif.

To demonstrate how to construct the unit motif, starting from the template motif we will pick the bottom left triangular sub-cell as our generator region. This is marked in black in Fig. 3.2b and the position of the template motif inside the chosen generator region is shown in Fig. 3.2c. Since the vertical boundary L1 of the generator region is a mirror line we can extend the motif by reflecting the template motif in L1 as shown in Fig. 3.2d. We can make the next extension by making use of the fact that **O** is a center of 3-fold rotation, this extension is shown in Fig. 3.2e. Finally by reflecting in the mirror line L2 we obtain the complete unit motif shown in Fig. 3.2f. All we have to do now is to copy the unit cell containing the unit motif to generate as much of the pattern as we care to make.

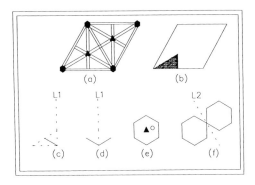

Figure 3.2:

A point to note is that the recipe described in Fig. 3.2 is not unique. We could have picked some other sub-cell as the generator region and made use of other centres of rotation and lines of reflection to construct the unit motif. There are many other possibilities, but the nature of the spatial relationships inside a unit cell is always the same for a pattern type. Figure 3.2a shows only part of the relationships for the type p6m. The complete set of relationships are shown in Fig. 3.15.

We have now indicated how symmetry analysis leads to the recognition of spatial relationships in a pattern and to minimal geometrical information needed to create a particular pattern. Is it not remarkable that the pattern in Fig. 3.1 can be reduced to just a tiny line segment? This example visually demonstrates the key feature of scientific activity. As Lewis Carroll might have said, science is about taking a grinning Cheshire cat and throwing away the cat to retain only the grin. Success in science comes from erasing as much of the mass of detail as possible and focusing only on the barest of essentials. Symmetry analysis of infinite patterns is an effective way to learn this lesson visually and in the context of art.

The reward for distilling the barest of essentials is that they can be

50   *Chapter 3. The Gateway from Islamic Patterns to Invariance and Groups*

connected in a natural and illuminating way with a large complex of things, which would otherwise appear to be diverse and unrelated. The reliance on great economy of thought in science gives science the facility to produce unity from diversity. We shall now demonstrate how through recognizing the symmetry properties of patterns we can learn to discover unity in patterns which appear quite different.

An unlimited number of p6m type patterns may be invented by merely altering the template motif. As to whether or not the inventions arising from such alterations will turn out to be appealing will depend on our imagination, experience, understanding of geometrical structure and possibly luck. The reader will recall, that in chapter 1 we pointed out that beautiful patterns can only arise if their is some scheme which provides for harmonious proportions. The interested reader may like to refer to a paper by Lalvani [23] who shows how known and new patterns can be generated through systematic subdivisions of the fundamental region.

For example, Fig. 3.3 shows a pattern obtained by replacing the straight line template motif of Fig. 3.1 with a circular arc. This pattern also occurs

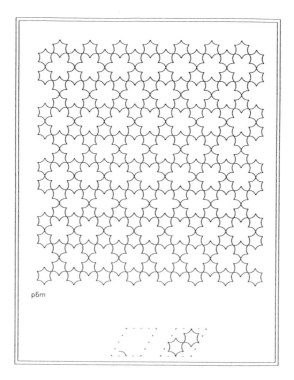

Figure 3.3:

3.1. An Example of Symmetries of an Islamic Pattern

on the first page of Bourgoin's book [6], where the suggested construction is of course quite different.

Figure 3.4 shows a pattern designed by us through making another very minor variation. There are infinite number of possibilities for producing variations on the theme.

In Fig. 3.5 we show how an interlaced version of Fig. 3.3 is produced by choosing the template motif as drawn in diagram **a**. The overlapping portions destroy the mirror lines in the pattern and what is produced

Figure 3.4:

52   Chapter 3. The Gateway from Islamic Patterns to Invariance and Groups

Figure 3.5:

is not a p6m type. In fact it becomes a p6 type pattern whose symmetries do not include any mirror reflections. Such a pattern has a different generator region, see Fig. 4.21. But we can, with a small amount of ingenuity, make this pattern by the same method as the one explained in Fig. 3.2. This can be done as follows.

A p6m type pattern will be produced if we use the unit cell **b** which is obtained from **a** by performing the same symmetry operations as those shown in Fig. 3.2. To produce the interlacing, we add to **b** the extra

3.1. An Example of Symmetries of an Islamic Pattern   53

lines shown in **c**. The unit cell **d**, obtained by combining **b**and **c**, when repeated, produces the interlaced pattern.

Finally, one more example in Fig. 3.6 is given to make the point that patterns which appear quite different and totally unrelated may in fact be based on identical spatial relationships. The patterns shown is also a p6m type pattern despite the fact that it appears to have no connection with the pattern in Fig. 3.1. The reader is invited to try and work out the template motif, the unit motif, and the unit cell for this pattern without looking at Fig. 3.7. where they are drawn.

Figure 3.6:

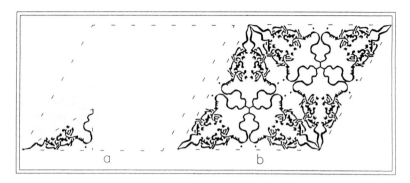

Figure 3.7:

3.2 THE KEY PROPERTY OF SYMMETRIC OBJECTS

Having given an outline of the method to be developed more fully in the next chapter, and having demonstrated some of its virtues, we will now develop the key concepts on which the method is based. The importance of these concepts will then be explained.

So far we have used the word symmetry without giving any definition. How can we define symmetry? The everyday definition of symmetry as an exact correspondence in size or shape between opposite sides of a structure, or regularity between parts, is not the one that proves to be the most powerful. To develop the definition of symmetry as used in science, we will examine two simple prototype symmetric objects.

3.2.1 Finite Symmetric Objects With Centres of Rotations and Lines of Mirror Reflections

Consider the square shown on the left in Fig. 3.8a. It has the property that if we rotate it about its center O by 90, 180, 270 degrees then although the vertices are permuted, the square **fits exactly in the spatial region that it did before we performed the rotation**. Assuming that there are no other distinguishing mark on it, it would seem not to have changed its position at all. If we rotate it by 360 degrees, then of course all the vertices return to their original positions and this rotation is the same as the one when we apply no action at all.

3.2. The Key Property of Symmetric Objects

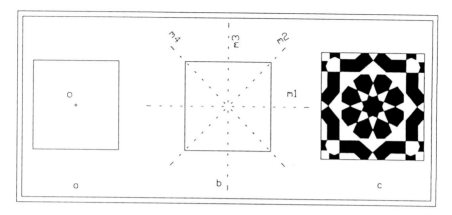

Figure 3.8:

Similarly, if we reflect it in any of the lines m1,m2,m3 or m4 which join the mid points of the sides or form the diagonals, as shown in Fig. 3.8b, then once again the appearance remains unchanged. We merely obtain a congruent copy, i.e., a copy which can be made to coincide exactly with the original. Symmetry of the square arises from these fundamental properties of the square.

Rotations about its center by multiples of 90 degrees, reflections in the lines joining the mid points of its sides, and reflections in its diagonals are called **the symmetries** of the square. Note that although the pattern in Fig. 3.8c looks more complicated than the unadorned squares on its left, its symmetries are identical.

Figure 3.9:

In finite symmetric objects, i.e., those which occupy some limited region of space, rotations, and reflections are the only kind of actions which can produce a congruent copy of the original. Several examples of such symmetric objects are shown in Fig. 3.9. In infinite objects, i.e., those that are assumed to extend indefinitely, there are two other types of transformations which can produce the same end result. These will now be examined.

3.2.2 Infinite Symmetric Objects with Translations and Lines of Glide Reflections

Examine the frieze pattern in Fig. 3.10. The pattern is assumed to extend indefinitely in both directions. Here, the distance between a point and its next repeat is L. Clearly, if we drag the pattern horizontally by a distance L or 2L or any integer multiple of L, then the pattern will look exactly the same as it did before.

Figure 3.10:

DEFINITION: A movement which causes every point in an object to shift by the same amount in the same direction is called a **translation**.

Any horizontal translation which is an integer multiple of L will not alter the appearance of this pattern. All such translations are symmetries of the pattern in Fig. 3.10.

Next, examine the footprint pattern in Fig. 3.11. If we apply a horizontal translation by 2L, or 4L, or any even integer multiple of L, then the pattern moves to a position of coincidence with its original location and does not change its appearance. In this case, however, we can perform another action which leaves the pattern unchanged. This involves a horizontal translation L followed by a reflection in the horizontal line passing through the middle of the pattern. The same holds if we translate by 3L, 5L or any odd integer multiple of L and then perform the reflection.

3.2. The Key Property of Symmetric Objects

Figure 3.11:

DEFINITION: A combination of a translation and a reflection is called a **glide reflection**.

We see that translations by any even multiple of L and glide reflections by any odd multiples of L are the symmetries of the footprint pattern.

The square and the foot-print pattern which we have just examined serve as prototypes for the two different types of spatial objects which display symmetry. Some objects are single entities like the square, but others such as friezes and wallpaper patterns are similar to the foot-print pattern, in being made up from repetitive elements.

3.2.3 Definition of Symmetry for Geometrical Objects

Having studied the prototype symmetric objects in the last section, we can now construct more general definitions, which apply to all symmetric geometric objects.

DEFINITION: An object is symmetric if there are translations, rotations, reflections, or glide reflections which when applied to the object leave the appearance of the object unchanged.

Mathematicians often express the above definition in an equivalent but slightly more technical language:

DEFINITION: An object is symmetric if the position in space occupied by it remains **invariant** to one or more **isometry** transformations.

By a **transformation**, we simply mean some action which causes an object to change its initial state.

The term **invariant**, as we shall see later, though appearing innocuous in meaning, gives rise to a concept of profound significance. The invariant properties of an object under a transformation are those properties which remain unaltered by the transformation.

An **isometry** transformation is a distance preserving transformation, i.e., a transformation which does not alter the distances between any two points in an object. Or to put it in our newly introduced terminology — distance is an invariant under an isometry transformation. It can be proved that there are just four such transformations — translations, rotations, reflections, and glide reflections.

Having developed a general definition of the adjective *symmetric*, which applies to all all geometric objects, we can now do the same for the noun *symmetry*.

DEFINITION: A transformation which has the property that when applied to an object, it leaves the appearance of the object unchanged is called a **symmetry** of that object.

3.2.4 Symmetry Group of a Geometric Object

Apart from symmetry, the other major concept to be introduced in this chapter is that of group. The term *group* has a special meaning in mathematics and its significance will be explained later. As a first step towards defining a group, we shall introduce the symmetry group of a geometric object.

The collection of all symmetries, i.e., all those rotations, reflections, translations, and glide reflections which, when applied to the object, cause the object to appear as if its position has remained unchanged, is called the **symmetry group** of the object.[1] Let us find the symmetry group of the square whose symmetries we examined earlier.

There are three distinct rotations around the center O which are symmetries. R_1, the rotation by 90^0; R_2, the rotation by 180^0; and R_3, the rotation by 270^0. The rotation by 360^0 has the same effect as rotation by 0^0, i.e., no rotation at all, and would seem to be of no consequence. However, it turns out that for theoretical purposes it plays a role as essential as the one played by zero in our number system. We refer to this trivial rotation as the **identity transformation**.

DEFINITION: An identity transformation, or more simply the **identity**, is a transformation which leaves every point of an object fixed in its original position. It is the "do nothing" or "stay put" transformation and is usually denoted by the letter I.

---

[1]Strictly speaking a symmetry group is not simply a collection or a set. It is the set together with a rule for combining the elements. To keep things simple, while discussing geometric objects, we will take it for granted that translations, rotations, reflections, and glide reflections can be combined.

3.2. The Key Property of Symmetric Objects

Apart from the identity and the 3 rotations, there are 4 mirror reflections which are symmetries of the square. Let us denote the reflection transformations in the lines m1, m2, m3, and m4 as M_1, M_2, M_3, and M_4, respectively.

Thus the symmetry group of the square is the set:

$$\{I, R_1, R_2, R_3, M_1, M_2, M_3, M_4\}.$$

This set is called the **dihedral group** D_4, the subscript denoting the fact that it comprises 4 rotations (including the identity) and 4 reflections.

The reader is invited to try evaluating and writing down the dihedral group D_3, i.e., the symmetry group of an equilateral triangle. There should be no difficulty after that in writing down the dihedral group D_n, comprising all the symmetries of a regular n sided polygon, where n is any positive integer.

3.2.5 The Four Special Properties of a Symmetry Group

To appreciate what is to come later, let us now show the key properties of the transformations which form the symmetry group of the square. First of all there is the obvious property that they can be **combined**, just as numbers can be combined via addition or multiplication. The rule for combining them is self evident. Thus a rotation of 90^0, followed by a rotation of 180^0 can be combined to produce a total rotation of 270^0, and similarly for any combination of rotations and reflections. Apart from the existence of a rule for combining the transformations, we note the following four relationships:

1. There is in identity transformation in the set. This of course is the one we have denoted with I. Its action is to do nothing.

2. No matter how many symmetry transformations we apply and no matter in which order we apply them, we always finish up with the square in a position which could have been reached by applying just one transformation from the group. This property is very easy to see for the rotations alone but it holds for any combination of transformations in the group.

   For example, if we rotate by $R_1 = 90^0$ and follow it up with a rotation of $R_2 = 180^0$, then this is the same as $R_3 = 270^0$. Let us write $R_2 R_1$ to mean the combined result of applying R_1 and following it up with R_2. Then $R_2 R_1 = R_3$.

Now, whereas this property is easy to see for rotations alone, it is not obvious if we mix reflections and rotations.

Examine Fig. 3.12a, which shows what happens to a point P in the square when we first reflect in the line m1 and then perform a rotation of 90^0 about O. Then the first operation sends P to P' and the second one sends P' to P''. But P can be moved to P'' by the single operation of reflecting in the line m2. Thus we see that $R_1 M_1 = M_2$.

The reader is invited to try some other examples to verify that the consecutive application of any two or more symmetry transformations from the symmetry group produces a transformation which is already included in the group. Mathematicians refer to this property as that of **closure**. We have shown examples to demonstrate that the symmetry group of the square is **closed** under the operation of combining transformations.

Figures 3.12b is intended to show that, in general, the order in which we perform the transformations is important. Here, we have reversed the order of the operations shown in the previous figure. First, the rotation of 90^0 sends P to P' and then a reflection in the line m1 sends P' to P'', showing that $M_1 R_1 = M_4$. Thus $R_1 M_1 \neq M_1 R_2$.

3. A third characteristic property of our symmetry group is that in combining three or more symmetry operations, it does not matter how we combine the members together, so long as we do not alter the order of the individual symmetries.

    Consider for example the combination $R_2 R_1 M_1$. Then this property implies that
    $$R_2(R_1 M_1) = (R_2 R_1) M_1.$$

    To appreciate what is being said, examine Figs. 3.12c and d. Since, as we saw earlier, $R_1 M_1 = M_2$, we can get rid of the bracketed terms and reduce the left hand side of the above equation to $R_2 M_2$. Figure 3.12c, shows the effect on a point P of applying the transformation $R_2 M_2$.

    Similarly, since $R_2 R_1 = R_3$, we can replace the right hand side of the above equation by $R_3 M_1$. Figure 3.12d shows the effect on a point P of applying $R_3 M_1$. We see that in both cases P finishes up at the same position P''.

    This property is called the property of **associativity** by mathematicians.

4. Finally, for every transformation in the symmetry group, there is one which un-does its effect. For example if we apply R_1, i.e., rotate by

90^0 and follow this by R_3, rotating by 270^0, then the net result is the same as that of no action. Thus R_3 is the inverse of R_1. Similarly, R_1 is the **inverse** of R_3. Note that any mirror reflection is its own inverse.

The combined action of a transformation and its inverse is equivalent to the action of the identity.

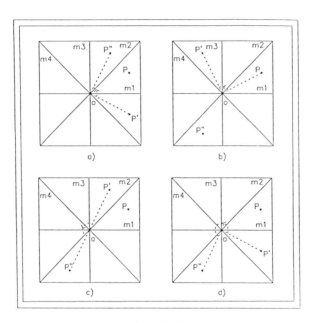

Figure 3.12:

3.3 THE SYMMETRY GROUP OF AN ISLAMIC PATTERN

Having seen the symmetry group of a simple finite object, we will now return to the Islamic pattern which we first examined in Fig. 3.1 and work out its symmetry group. The pattern has an infinite number of symmetries. We will show how the symmetry group can be depicted graphically in this case when it comprises an infinite collection.

Before continuing with the rest of this section the reader is asked to construct a simple gadget in the form of a transparent overlay. For this, find a piece of a transparent sheet; for example, the kind used for overhead projectors would be ideal. Cut it approximately to the size of Fig. 3.13c. Place the sheet securely on Fig. 3.13c and using a black felt-tip pen copy the star-hexagon pattern shown in solid black lines.

As we go through this section, the reader is invited to employ the transparent overlay for verifying symmetries. To do this, place the overlay on top of one of the figures in Fig. 3.13, bringing the pattern on it into coincidence with the pattern below. Now apply to the overlay the appropriate movement for the symmetry transformation to be examined. In the resulting position, the pattern on the overlay will again coincide with the pattern below.

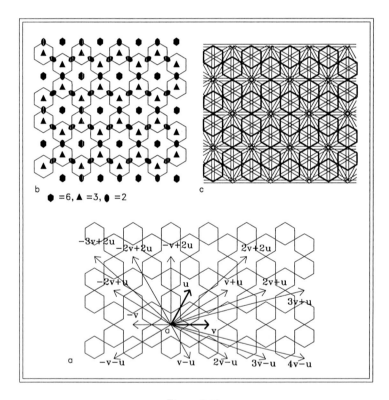

Figure 3.13:

3.3.1 The Translation Symmetries

The most basic symmetries of a repeat pattern are the translation symmetries. Every such pattern, because of its very nature of construction, must have these. What this means is that there are to be found an infinite number of different directions along which we can slide the pattern, to make it coincide with itself. Some of these are marked with arrows in Fig. 3.13a.

In each of these directions, there is a minimal translation. All integer multiples of this translation are also symmetries of the pattern. Amongst this infinite number of translations, there are two that have the smallest magnitudes. These are indicated with bold arrows and marked **u and v**. These two minimal translations are said to form **a basis** for all other translations. Any other translation symmetry transformation can be achieved by successive applications of **u, v** as shown in the diagram. The reader is invited to verify these using the transparency.

3.3.2 The Rotation Symmetries

Figure 3.13b shows all the centres of rotations about which the pattern exhibits rotation symmetries. There are to be found 6-fold, 3-fold, and 2-fold centres of rotations and these are marked with different symbols as explained in the diagram.

The 6-fold centres of rotations were already pointed out in our initial discussion. The center of every star shape has this property. It is easily seen that the center of every hexagon shape is a center of 3-fold rotation.

Whereas the 6-fold and the 3-fold centres are fairly obvious, the 2-fold centres are not so. The reader may like to employ the transparent overlay to verify the 2-fold rotation symmetries. Place the transparency to coincide with Fig. 3.13b, then place a pencil point or some such object to hold down a 2-fold center. Now rotate the overlay by 180^0. It will be found that the patterns in the two figures come into coincidence.

To achieve a 180^0 rotation the reader will find it useful to draw, using a washable felt tip pen, a short line segment on the transparency to coincide with a line segment on the pattern below. When the turn is complete the two segments will line up.

Comment: (i) The only rotation symmetries that can occur in infinite repeat patterns are 2-fold, 3-fold, 4-fold, and 6-fold. This is often referred to as the **crystallographic restriction**.

(ii) The most likely positions where centres of rotations of a pattern may occur are centres of polygons and star shapes. The other likely candidates are vertices and midpoints of sides.

3.3.3 The Mirror Reflection Symmetries

Figure 3.13c shows all the mirror lines about which the pattern exhibits reflection symmetries. There are six sets of mirror lines passing through every 6-fold center of rotation and again we met these in our initial discussion of this pattern.

As suggested for rotation symmetries, the reader may like to employ the transparent overlay to verify the mirror symmetries. To check for a line of mirror symmetry, place the overlay on top of the figure as before. On the transparency, mark the position of the mirror line of interest with two points. Now flip the overlay over the mirror line, using the points as guides to its position. Verify that the pattern on the overlay coincides with the pattern below.

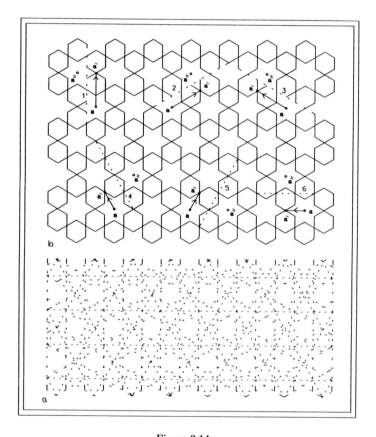

Figure 3.14:

3.3.4 The Glide Reflection Symmetries

Just as there are six sets of mirror lines, so there are six sets of glide lines in the pattern. These are shown in the bottom diagram in Fig. 3.14a. If we were to draw all of them on the pattern then the pattern will be completely obscured. In Fig. 3.14b we have drawn a small portion of one glide line from each of the six infinite sets. These are shown dotted.

Now examine Fig. 3.14b. Consider the point marked **a** at the top left. The arrow from **a** to **a′** represents half the translation necessary to make the pattern coincide with itself. If we first translate the pattern by this amount, so that the point **a** moves to **a′**, and follow this by a reflection in the dotted line then it is easy to see that **a′** will move to **a″**. Now **a″** is the center of a star shape exactly as is **a**. Through these two actions, involving a translation and a reflection, not only the point **a** but every point in the pattern moves to a correspondingly similar point on the pattern. Exactly the same holds for the other five glide lines shown in the same diagram.

Again the reader may like to employ the transparent overlay to verify the glide symmetries. To do this, place the overlay on top of Fig. 3.14b as before. On the transparency, mark the position of the glide line of interest with two points. Also mark the positions of **a** and **a′**. Now slide the transparency to cause **a** to move to **a′** and flip the transparency over the dotted glide line. Verify that the pattern on the overlay coincides with the pattern below.

3.3.5 Symmetries Depicted in a Unit Cell

We have now examined all the symmetries of our pattern and we see that there are an infinite number of them. This infinite collection is the symmetry group of the pattern. This infinite set can be listed symbolically using algebraic notation but a much more convenient way of depicting it pictorially is shown in Fig. 3.15.

This figure is a parallelogram whose sides are the minimal translation symmetries **u, v**. This parallelogram, as mentioned earlier, is usually called a **unit cell** but is also referred to as a **lattice unit** in some mathematical literature and as a **repeat unit** by pattern designers. The infinite symmetry group is condensed by marking all the symmetries on the unit cell as shown in Fig. 3.15, where the notation used is exactly the same as the one used earlier in Fig. 3.13 and Fig. 3.14. This diagram is common to all p6m type patterns and we shall see it again in the next chapter.

The complete symmetry group can be generated by successive translations **u, v** and copying of the condensed one marked on the unit cell in

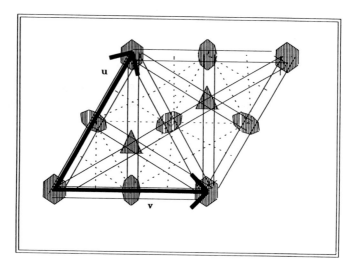

Figure 3.15:

Fig. 3.15. Any extent of the pattern can be generated by the same translations and the copying of the design contained in the cell.

3.4 SYMMETRY AND GROUPS IN GENERAL

Appearing to remain unchanged when acted on in some way is not a property which arises only in the context of symmetric geometrical patterns or spatial structures. It occurs with astonishing frequency in natural phenomena as well as in abstract knowledge. Let us now give some simple non-geometric examples of the occurrence of invariance.

Example 1: $x^2 + y^2$

The expression is symmetrical in x and y. Here, if we change x to $-x$ or y to $-y$ the expression does not change. It remains invariant under the action of negation. Also, if we change x with y and y with x, i.e., if we permute x and y, then this action also leaves the expression invariant. The actions of negation and permutation are, by analogy with the geometric case studied above, the symmetries of this expression.

Example 2: $x^2y - xy^2 - x^2z + xz^2 + y^2z - yz^2$

3.4. Symmetry and Groups in General

This expression has a more constrained symmetry than the one in Example 1. In general, if we permute x, y, and z then it is altered. But it does remain invariant if we make a cyclic permutation in x, y, and z, i.e., if we replace x with y, y with z, and z by x, or x by z, z by y, and y by x. Thus the cyclic permutation is a symmetry of this expression.

These examples, though simple, are not pointless. The kind of algebraic symmetries we have just pointed out in Examples 1 and 2 above form the basis of the theory of equations of degrees higher than four.

Example 3: A theorem in simple geometry says that if a triangle has two sides equal then the angles opposite to the two sides are also equal.

If we interchange the words 'sides' and 'angles' then the resulting theorem is also true. Thus, in this example, the truth of the theorem remains invariant under the transformation which interchanges the words sides and angles. This transformation is thus a symmetry of the theorem.

It is useful to remark that the symmetries of this elementary theorem reflect on the nature of the Euclidean space. If we were drawing triangles on the surface of a sausage-shaped balloon, say, then the theorem will not hold.

Example 4: Faraday's law: A changing magnetic field gives rise to an electric field. If we interchange the words 'electric' and 'magnetic' then the law so obtained still holds. Here we have a law of nature whose truth remains invariant under the transformation which interchanges the words 'electric' and 'magnetic'.

Example 5: Moral imperative: Do unto others as you would have others do unto you. The symmetry is obvious.

Thus we see that there can be all kinds of systems which display this attribute that some of their properties remain *invariant to change* under some kind of transformations. They posses symmetry. It is also true that in general the symmetry property in theorems, results, scientific laws, physical structures, and other systems can be utilized to gain insight about the nature of what is involved. For this reason we are led to formulate the following general definition of symmetry.

Definition: A system is symmetric if there are one or more transformations which when applied to the system leave some attribute of the system invariant. Such a transformation is called a symmetry of the system.

It is also true that there are to be found many kinds of systems, geometric, physical, and abstract whose elements, when combined by some rule, satisfy the same four properties that were seen to be satisfied by the eight elements of the symmetry group of a square, namely, **closure, associativity, the existence of an identity, and the existence of inverses**. Let us again look at some simple examples.

Example 6: Consider the set of integers and take the rule for combining them to be the ordinary everyday process of addition. Since adding two integers produces another integer, the condition of closure is clearly satisfied. That associativity holds may be seen equally easily from an example, e.g., $(2+3)+4 = 2+(3+4)$. Note that if we choose our rule for combining elements to be subtraction instead of addition, then associativity will not hold, since $(2-3)-4 = -5$, but $2-(3-4) = 3$. The number zero serves as the identity, since adding zero to an integer produces no change in it, $4+0 = 0+4$, etc. Finally adding the negative of any integer to itself produces the identity zero, $4+(-4) = 0$, etc. Hence, with this rule, every element has an inverse.

Example 7: Suppose that you are taking a walk on a road. To make things simple, let the road be straight and extend indefinitely in both directions. Suppose that you walk from a position O to a position P and then walk from P to a position Q. Let OP mean the action of walking from O to P and + mean the action of combing two walks. Then the combined action OP+PQ=OQ. Thus the walks satisfy closure. $(OP+PQ)+QS = OS$ and $OP+(PQ+QS) = OP+PS = OS$, hence we have associativity. Standing still serves as the identity and for every element OP, the element PO serves as the inverse. Thus all the four properties are satisfied.

It is worth remarking that although the last example seems trivial, it is in fact the basis of a powerful theory in Physics known as Lie Groups, named after the Norwegian mathematician Sophus Lie, who initiated it during the 1870s.

We are now motivated to formulate a general definition to describe systems of the kind that we have just examined in the last two examples. Such a system is called a **group** and is defined as follows:

Definition: Let G be a system comprising a set of elements (which we will denote by a, b, c, and so on), together with a rule of combination (which we will denote by the symbol $\oplus$). G is called a **group**, if and only if its elements satisfy the following four conditions:

(i) **Closure:** If a and b are elements of G, then $a \oplus b$ is also an element of G.

(ii) **Associativity:** If a, b, c are elements of G, then $(a \oplus b) \oplus c = a \oplus (b \oplus c)$.

(iii) **Identity:** G contains an element I (called the identity) such that if a be an element of G, then $a \oplus I = I \oplus a = a$.

(iv) **Inverse:** For element a in G, there exists another element a^{-1}, such that $a \oplus a^{-1} = a^{-1} \oplus a = I$.

The study of properties of groups has given rise to the branch of mathematics known as group theory. The foundations of the theory were laid in 1829, by a young French mathematician named Evariste Galois. His is a sad but romantic tale. He wrote his masterly paper on the subject at the young age of twenty. He was induced to fight a duel with a rival over his association with a beautiful woman. Alas he lost the duel.

Galois' worked was picked up and developed vigorously by a number of leading mathematicians of the nineteenth century. Group theory has found wide applications and turned out to be an invaluable tool in describing nature. E. T. Bell [4] remarked, "Wherever groups disclosed themselves, or could be introduced, simplicity crystallised out of comparative chaos."

3.5 THE TWO GRAND QUESTIONS AND ONE THAT IS GRANDEST

We have developed above the core concepts in symmetry and groups. The importance of these concepts arises from the fact that they point simply and directly to two of the most profound questions that may be asked about a variety of systems, both physical and abstract.

And what exactly are the questions? They are these:

1. Given a group of transformations, what aspects of a given system remain invariant, or immune to change, under their actions?

2. Given some aspect or aspects which must remain invariant, what groups of transformations can we discover, that leave them unaltered?

Physicist are led by these questions in their efforts to unravel the fundamental forces and particles of the phenomenal world. Chemists rely on answers to the same questions to explain the electromagnetic spectra of molecules. The same two questions serve topologists in studies of surfaces in higher dimensions. In just over a hundred years these two key questions have brought about a remarkable unification of algebra, geometry, and physics and have shed light on many other fields of human endeavor.

It is scarcely possible to emphasize the utility and power of these questions. What we shall attempt to do is just to trace the overview of the major revolutions that have resulted from the asking of these questions.

The first event that dramatically brought home the importance of these questions occurred in 1872, at Erlangen in Germany. There, a twenty-three year old mathematician named Felix Klien proposed a revolutionary new *group invariant* view of geometry. He defined geometry as "the study of those properties of a set that remain invariant when the elements of the set are subjected to the transformations of some transformation group." This view of geometry led to a fundamental revision in approach and restored a unified view of the subject which had become severely fragmented by the mid-nineteenth century through the birth of several non-Euclidean geometries. This revolutionary event in the history of mathematics has become known as the **Erlangen** or **Erlanger programme** and has left a permanent imprint on the growth of mathematics.

A similar revolution to the one unleashed in geometry by the Erlangen programme was initiated in Physics in 1905, when Albert Einstein published his famous paper on special relativity. Before Einstein's paper the concept of symmetry was not at all emphasized in physics. The laws of physics were talked about as **conservation laws**.

Typically, Newton's first law says that an object moves in a straight line at a constant velocity until acted on by some force. This old way of thinking about this law is in terms of conservation of momentum (mass X velocity) — in the absence of a force momentum is conserved. We can put it another way by saying that in the absence of a force the velocity remains invariant when a body is subjected to the transformations of translations and passage of time. We can see that this implies translations and the passage of time are symmetries of the system. Thus symmetries correspond to conservation laws and vice versa.

What Einstein did was to place the emphasis on symmetry. He abandoned the previously held absolute concepts of space and time, but made crucial the requirement that the laws of physics be invariant to two observers moving with constant velocity relative to each other. He emphasized the symmetry of physical laws with respect to certain transformations, known as the **Lorentz group**.

This approach which focuses on invariance has turned out to be more fundamental and has led to the concepts of group and symmetry becoming the most basic concepts in all modern theories of physics. It unites physics and geometry because both are seen as just group invariant theories and the change in philosophical point of view is often referred to as the "geometrisation of physics."

Even more basic than that, it has been pointed out by Rosen [41] that reproducibility and predictability, which are at the very foundation of the scientific method, are based on symmetry. This is so because science relies on the fact that the result of an experiment remains invariant no matter who performs them or where they are performed.

Indeed, in our quest for unity we can go even further and show that in a broader sense the significance of the two grand questions extends beyond physics. This was explained beautifully by C. J. Keyser in a lecture entitled *The Group Concept*, [22]. He said,

> "The sovereign impulse of man is to find the answer to question: what abides? ... Thought, — taken in the widest sense to embrace art, philosophy, religion, science, taken in their widest sense, — is the quest of invariance in a fluctuant world."

We could not agree more. It is the impulse to seek out invariance that truly drives all the most noble human enterprises and God can be thought of, in the widest sense, as the ultimate invariant. As the mathematician poet Omar Khyam put it:

Whose secret Presence, through Creation's veins
Running Quicksilver-like eludes your pains;
Taking all shape from Mah to Mahi; and
They Change and Perish all — but He remains.

The Quran offers the very same attribute for God:

Allahu Baqi min kulle Fani
— God remains all else is transient

3.6 CONCLUDING REMARKS

What we have done is to show that the studies of symmetry in the geometrical patterns of Islamic art leads naturally to the broader concept of symmetry and group theory, two of the supreme flowerings of the human scientific intellect. The central concept which underpins these great achievements is that of invariance and in the broader sense the search for invariance unites religion, art, philosophy, and science in a very profound way.

We conclude this chapter by quoting a line from Rumi:

Example seekest of science springing in the heart?

The message of this chapter may be put in condensed form as:

Then seek no more study Islamic art!

Chapter 4

CLASSIFICATION, IDENTIFICATION AND CONSTRUCTION OF THE SEVENTEEN TYPES OF TWO-DIMENSIONAL PERIODIC PATTERNS

Mathematics, which most people think of as dealing solely with number, is in fact the science of structure and pattern.

R. Buckminster Fuller

Sherlock Holmes, the pipe-smoking super-detective created by the English author Sir Arthur Conan Doyle, we are told, wrote *a little monograph on the ashes of one hundred and forty different varieties of pipe, cigar and cigarette tobacco.* Our task in this chapter, luckily, will not be anywhere near as daunting or as subtle as that undertaken by Holmes.

We shall be concerned in this chapter with the seventeen types of symmetric repeat patterns which can arise in two-dimensions. It will be explained how these patterns are classified and how they can be recognized from their symmetry properties. We shall also give simple algorithms, developed by the author, for their construction.

4.1 CLASSIFICATION OF PATTERNS

Mathematicians classify patterns according to their symmetry groups, which, as was explained in the last chapter, means the set of symmetries they possess. It needs to be emphasized again, however, that the term *group* implies the existence of a rule for combining the elements forming the group.

By identifying the symmetries of a pattern, we can identify its symmetry group and hence its type. The same symmetries can be utilized to construct the pattern in an efficient manner. All this was illustrated by an example in the last chapter for just one type of pattern, namely, the type p6mm. We shall now examine all the possibilities.

A two-dimensional periodic or repeat pattern, also called a **wallpaper pattern**, is produced when a motif of some sort is copied onto the nodes of a net constructed from two sets of parallel lines. Such a net is shown in Fig. 4.1. Thus, underlying a repeat pattern there is always a **lattice** or a **net**. The intersections of these parallel lines cover the plane with parallelogram shaped tiles. Any such tile forms a unit cell.

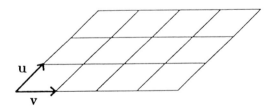

Figure 4.1:

It can be proved mathematically, see for example Martin [31], that if we enumerate all the distinct symmetry groups that are possible for two-dimensional repeat patterns then this number is seventeen. Hence, if we choose to classify repeat patterns by their symmetry groups, then seventeen and only seventeen different pattern types are possible.

Translations with any whole number of combinations of **u, v** are of course necessarily two symmetries of any repeat pattern. What other kind

4.1. Classification of Patterns

of symmetries may arise depends on the shape of the unit cell produced by the lattice. There are five different possibilities, arising from the fact that parallelograms fall into five different categories. These net types are shown in Fig. 4.2 and the possible symmetries that they can support are described below.

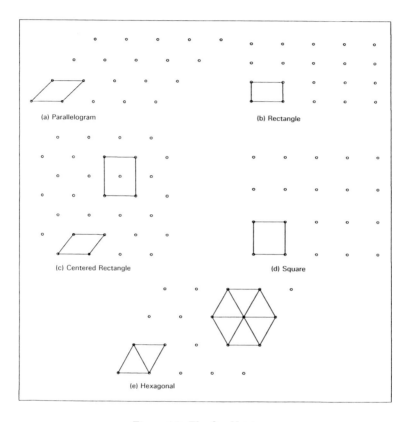

Figure 4.2: The five Net types.

4.1.1 The Five Net Types

The two sets of parallel lines which define the underlying net can give rise to the following five types of parallelograms for their unit cells:

1. **An Oblique Parallelogram With Unequal Adjacent Sides**
   Only 180^0 rotations, also called 1/2 turns, are the possible symmetries.

2. **A Rectangle**
   Mirror reflections in one or both sides and glide reflections are possible symmetries. Since a rectangle is a special case of a general parallelogram, we can deduce that 1/2 turns will be possible. For the same reason, 1/2 turns can arise on any net and from now on we shall take the existence of this symmetry for granted in every case.

3. **A Rhombus With Adjacent Sides Equal And Not Containing a 60^0 Angle**
   This type of net is referred to as a **centered rectangular net**. The name arises from the fact that since the diagonals of a rhombus bisect each other at right angles, the nodes of this type of net can be seen to form **centered rectangles**, as shown in Fig. 4.2c. This observation proves useful in identifying as well as in constructing patterns based on this net. Possible symmetries are mirror reflections.

4. **A Square**
   This kind of net allows 90^0 or 4-fold rotations, also called 1/4 turns. It can support mirror reflections at 90^0 as well as 45^0 to the cell sides.

5. **A Rhombus With Adjacent Sides Equal And Containing a 60^0 Angle**
   This special rhombus divides into two equilateral triangles, as shown in Fig. 4.2e. Six such triangles surrounding any node on the net can be grouped together to form a hexagon and the net can thus be seen alternatively as a tiling with hexagons. For these reasons, this special rhombic net is referred to as a **hexagonal** net.

   The net allows 60^0 (6-fold or 1/6 turn) as well as 120^0 (3-fold or 1/3 turn) rotational symmetries. It can support mirror and glide reflections at multiples of 60^0 to the cell sides.

4.1.2 The Seventeen Pattern Types

The seventeen possible types of two-dimensional repeat patterns are denoted by symbolic names, using up to 4 symbols, in an internationally agreed notation. This notation was invented by crystallogrophers. There are a number of other notations, but presently this is the one that is in use most widely. The names of the seventeen pattern types and the type of net on which they occur are listed in Table 4.1. The meaning of the notation is explained in the next section.

4.1. Classification of Patterns 77

Table 4.1: Table showing the symbolic notation used to denote the various pattern types supported by each of the five type of nets. A short form of the notation is often used for some of the patterns. Where this applies, the short notation is shown in brackets.

Unit Cell Shape	Pattern Types
Parallelogram	p1, p2
Rectangle	p1m (pm), p1g (pg), p2mm (pmm), p2mg (pmg), p2gg (pgg)
Rhombus	c1m (cm), c2mm (cmm)
Square	p4, p4mm (p4m), p4gm (p4g)
Hexagonal	p3, p3m1, p31m, p6, p6mm (p6m)

4.1.3 The International Crystallographic Notation

To explain the meaning of the symbols used to name the seventeen patterns listed in Table 4.1, it is necessary first to introduce some more terminology.

In the scheme of classification, we need to identify a unit cell which can generate the whole pattern by repeated translations only. If the unit cell that is used is the basic parallelogram cell generated by the net as shown in Fig. 4.1, then the cell is called a **primitive cell**. The primitive cell contains the smallest area from which a repeat pattern can be generated using translations only. Of course, we can generate a pattern by repeating a region containing several primitive cells, say four basic parallelograms. In that case, the repeat region will not constitute a primitive cell.

In two types of patterns, c1m and c2mm, both of which arise on a rhombic net, it is more convenient to use a rectangular cell rather than the primitive rhombus cell to be able to categorize their symmetries in a manner consistent with other types of patterns. This chosen rectangular repeat region is **centered** over the primitive rhombus and contains twice the area of the rhombus. Such a cell, for obvious reason, is called a **centered cell**. The centered cell is preferred because it makes the mirror lines, which arise in the cm and cmm patterns, parallel to the cell boundaries. This will not happen if the primitive rhombus cell was to be used.

Figure 4.3 shows a pattern whose primitive cell is a rhombus. A primitive cell together with a surrounding centered cell is shown. Note how the mirror lines in the pattern, shown as double lines, line up with the boundaries of the centered cell but not with the boundaries of the primitive cell.

We can now explain the symbols that occur in Table 4.1. There are two notations in common usage. One is a full notation and there is also a shortened version which is shown in brackets in Table 4.1. In general, in the full notation, four contiguous symbols of the form S1-S2-S3-S4 are utilized.

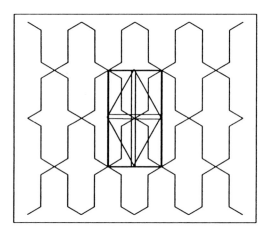

Figure 4.3:

But the last two symbols are dropped in some cases as explained below. The symbols that are utilized in the four positions and their meanings are as follows.

- The first symbol S1 is a 'c' if the net is centered rectangular; otherwise it is 'p' (for a primitive cell).

- The second symbol S2 denotes the highest order of rotation symmetry in the pattern. For example, a '3' in this position would indicate that 120^0 rotations are the highest. If the second symbol S2 is a '1' then this indicates that no rotational symmetry is present. The symbol S2 can take one of the values '1', '2', '3', '4', or '6'.

- The third symbol S3 is an 'm' (for mirror reflection) if there is a mirror reflection line. It is a 'g' if there is no mirror reflection, but there is a glide reflection. The symbol S3 is a '1' if neither mirror or glide symmetries are present.

- The symbol S4 is used in the same way as the symbol S3, to indicate the presence of a mirror line or a glide line in a second direction.

Note: The third and fourth symbols are ignored if there are no mirror reflections or glide reflections.

Also, apart from the cases p3m1 and p31m, the four symbol notation can be shortened without causing any ambiguity in identification. For example from c2mm, the 2 is dropped and the notation is shortened to cmm. The 'mm' in the third and fourth place imply the existence of 2-fold rotations and hence '2' can be dropped without any loss. For these reasons, the short notation is the one that is used more commonly.

4.2 EXAMPLES OF THE SEVENTEEN TYPES OF PATTERNS FROM ISLAMIC ART

It is now time to look at all the seventeen types of patterns and since we are concerned with Islamic patterns, we shall take our examples from them. Several other authors have examined Islamic patterns for their symmetries and we shall give a brief account of their work in the next chapter. Examples chosen by us are drawn in Figs. 4.6–4.22. and photographs of actual sources corresponding to the figures are shown in Photos 4.1–4.17. The reader is invited to now examine the diagrams and the photos, but it would be advisable to read through simultaneously the explanations and comments that follow in the rest of this section.

In each of the Figs. 4.6–4.22 there are to be found four smaller diagrams below the main pattern. These show the following:

The lower left diagram shows the general shape of the unit cell for the pattern type and the dark shaded portion shows the **fundamental region** or the **generator region**. We met these terms in the last chapter for a p6m type pattern. In the general case they are defined as follows:

Definition: A fundamental region (or a generator region) for a given pattern type is a smallest region in area from which the entire pattern can be generated by the action of the symmetry transformations in the symmetry group of that pattern type.

Such a region contains all the information needed to generate the complete pattern. The idea is quite simple and was illustrated in the last chapter for a p6m type pattern. To recapitulate, this is what we do.

We place a motif in the generator region. This motif we have called the **template motif**. We apply suitable symmetry transformations from the symmetry group to generate the motif in the complete unit cell. We have called this motif a **unit motif**. We then generate the entire pattern repeated translations of the unit cell on the lattice and the copying of the unit motif contained in it.

Going back to the diagrams under discussion, the unit cell is again drawn at the bottom right but this time it shows all the symmetries, i.e., the

80   Chapter 4. Classification, Identification and Construction of ...

Photo 4.1:

Photo 4.2:

4.2. Examples of the Seventeen Types of Patterns from Islamic Art

Photo 4.3:

Photo 4.4:

82   Chapter 4. Classification, Identification and Construction of ...

Photo 4.5:

Photo 4.6:

4.2. *Examples of the Seventeen Types of Patterns from Islamic Art*   83

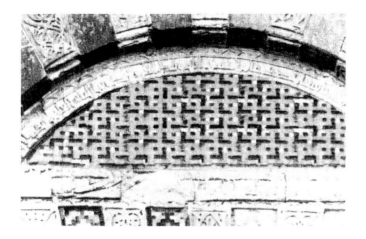

Photo 4.7:

Photo 4.8:

84   Chapter 4. Classification, Identification and Construction of ...

Photo 4.9:

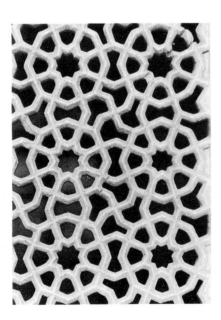

Photo 4.10:

4.2. Examples of the Seventeen Types of Patterns from Islamic Art   85

Photo 4.11:

Photo 4.12:

86 Chapter 4. Classification, Identification and Construction of ...

Photo 4.13:

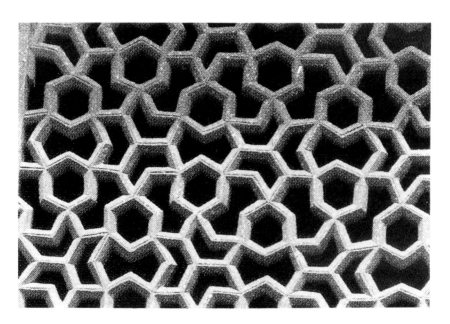

Photo 4.14:

4.2. *Examples of the Seventeen Types of Patterns from Islamic Art* 87

Photo 4.15:

Photo 4.16:

88   *Chapter 4. Classification, Identification and Construction of ...*

Photo 4.17:

symmetry group, of that pattern type. Since translations are in every symmetry group, these have not been marked. This type of diagram was developed and its meaning was fully explained in the last chapter; see Fig. 3.15. The notation used to denote various symmetries is given in Fig. 4.4. and the diagrams are based on a paper by Schattschneider [43] where the reader will find more details.

The two small diagrams in the middle show the template motif and the unit motif used to generate the particular pattern shown on the top.

At this stage there is a point worth making, which the reader may not have fully appreciated. With reference to Table 4.1 it should be understood, for example, that whereas p1 and p2 type patterns can be constructed on a general parallelogram net, they can also be constructed on any of the other types of nets. This is so because the shapes rectangle, rhombus, square, or the hexagonal rhombus are also parallelograms. For example, Fig. 4.6 shows a p1 type pattern. In the most general case such a pattern has an arbitrary parallelogram for a unit cell. In this particular example the unit cell is a rectangle.

Similarly, any pattern type that can occur on a rhombic net may also occur on a square net, since a square is also a rhombus. On the other hand, a p4 type cannot occur on a rectangle because a rectangle is not a square. Thus Table 4.1 must be interpreted in this way. Figure 4.5 is drawn to further emphasize this point.

4.2. Examples of the Seventeen Types of Patterns from Islamic Art

symbol	meaning
○	Center of 2-fold rotation
△	Center of 3-fold rotation
□	Center of 4-fold rotation
⬡	Center of 6-fold rotation
═	Line of mirror reflection
- -	Line of glide reflection

Figure 4.4: Notation used to indicate various types of symmetries.

In Fig. 4.5a we have a p6m type pattern which can only occur on a hexagonal net. By introducing a motif element which destroys the p6m symmetry we have produced a p1 type pattern on a hexagonal net which is shown in Fig. 4.5b. This observation leads us to introduce the concept of **ground symmetry**.

In examining the symmetries of a pattern we should take into consideration all its features such as colors, inscriptions, interlacing, and other adornments. Thus, if we claim that a particular line is a mirror line then every such feature should exactly reflect in it, and so on. When the extra features are ignored and we concentrate only on the geometrical lines or curves which are assumed to be infinitesimally thin, then we say that we are examining the pattern in its **ground symmetry**.

Thus the pattern in Fig. 4.5b is of the type p1 but its ground symmetry is p6m. The reader should keep this distinction in mind when comparing the diagrams with the photos. In some of the photographs, the pattern is only of the type shown in the corresponding diagram if we consider its ground symmetry. The confusion between symmetry and ground symmetry has often been a source of controversy in the literature. In comparing the photos in this chapter with the corresponding patterns drawn in Figs. 4.6–4.22, the reader should keep this in mind.

Another point to appreciate is that knowing the symmetry type of a pattern does not necessarily supply us with useful information about the aesthetic appeal of the pattern. This is demonstrated in Fig. 4.5c. It shows

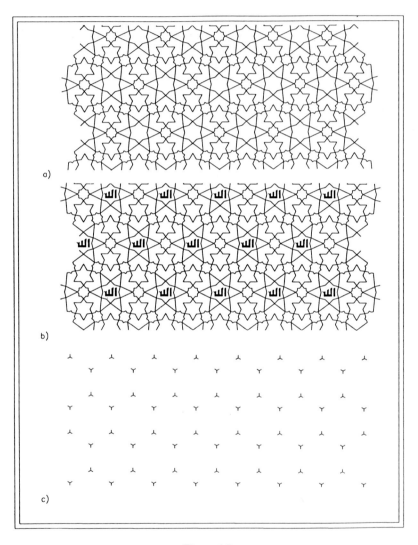

Figure 4.5:

4.2. *Examples of the Seventeen Types of Patterns from Islamic Art* 91

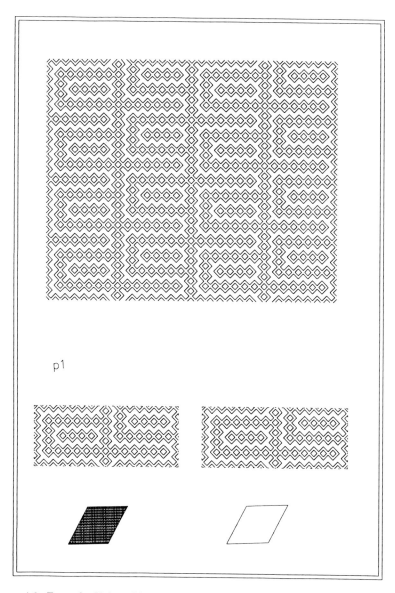

Figure 4.6: From the Kalyan Minaret, Bukhara, Uzbekistan; early 12th century (after E. Makovicky [28]). Photo 4.1.

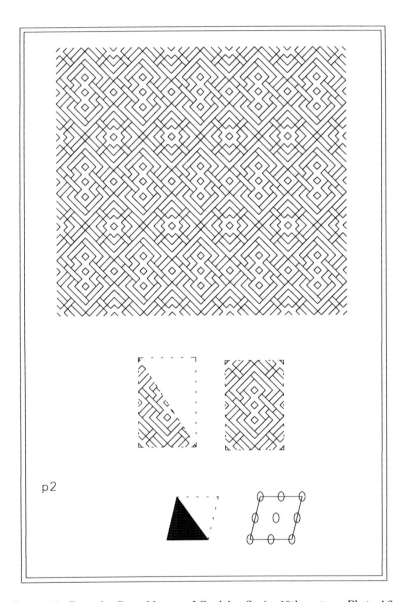

Figure 4.7: From the Great Mosque of Cordoba, Spain; 10th century. Photo 4.2.

4.2. Examples of the Seventeen Types of Patterns from Islamic Art 93

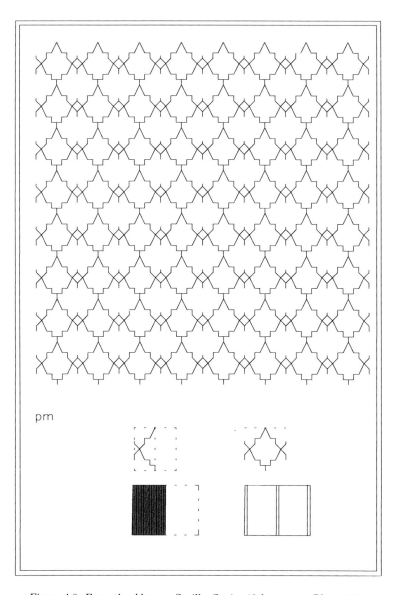

Figure 4.8: From the Alcazar, Seville, Spain; 12th century. Photo 4.3.

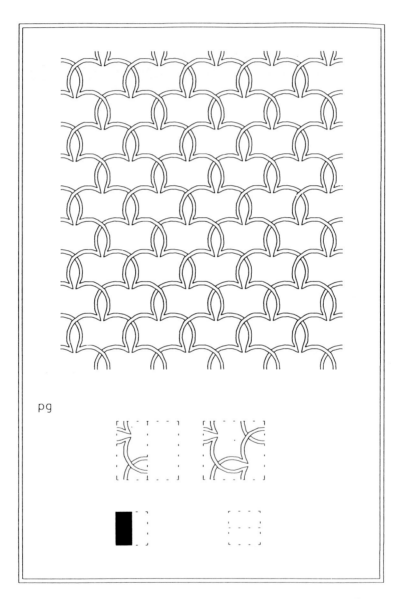

Figure 4.9: From the Alhambra, Granada, Spain; 14th century. Photo 4.4 (after Montesinos [32]).

4.2. Examples of the Seventeen Types of Patterns from Islamic Art

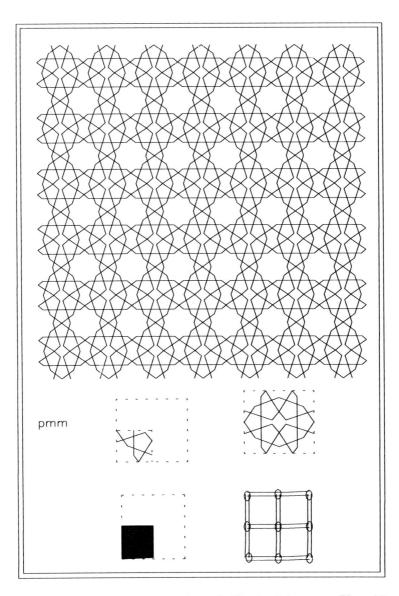

Figure 4.10: From the mausoleum of Itmad-al-Daula, 17th century. Photo 4.5.

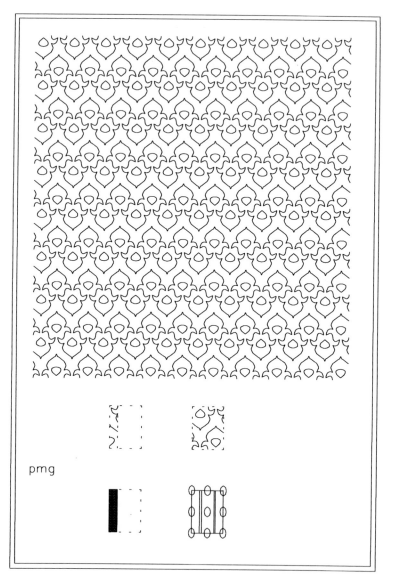

Figure 4.11: From the El-Muayyad Mosque, Cairo, Egypt; early 15th century (after Bixler [5]). Photo 4.6.

4.2. Examples of the Seventeen Types of Patterns from Islamic Art   97

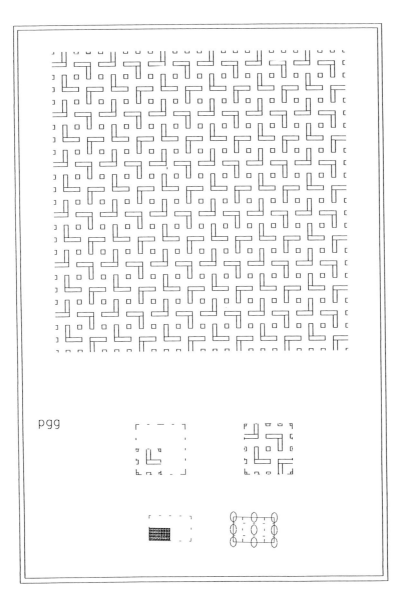

Figure 4.12: From the Great Mosque of Cordoba, Spain; 10th century. Photo 4.7.

98   Chapter 4. Classification, Identification and Construction of ...

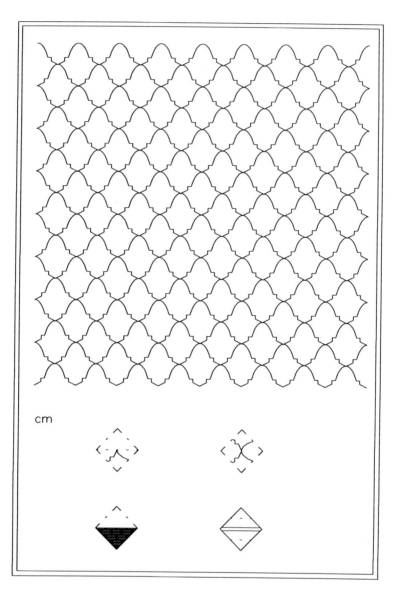

Figure 4.13: A frequently occurring motif in Islamic buildings in Spain and Morocco, known as *ktaf u darj* — *shoulder and step*. Photo 4.8 is from Bab Mansour, Meknes, Morocco; 17th century.

4.2. *Examples of the Seventeen Types of Patterns from Islamic Art* 99

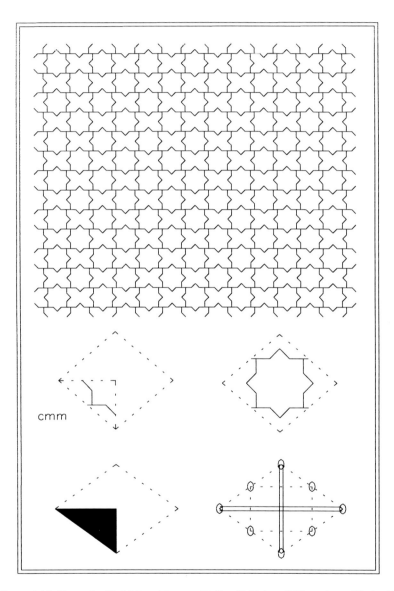

Figure 4.14: From the Shahjahan Mosque, Tatha, Pakistan; 17th century. Photo 4.9.

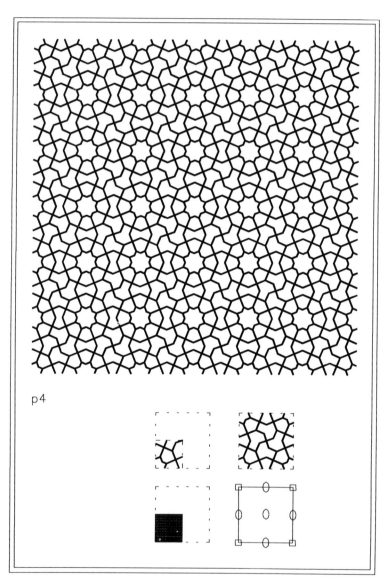

Figure 4.15: From the Tomb of Shaikh Salim Chisti, Fatehpur Sikri, India; 16th century. Photo 4.10.

4.2. *Examples of the Seventeen Types of Patterns from Islamic Art* 101

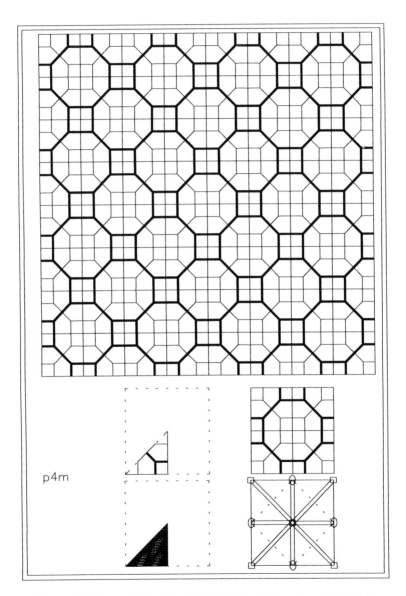

Figure 4.16: From the Red Fort, Delhi, India; 17th century. Photo 4.11.

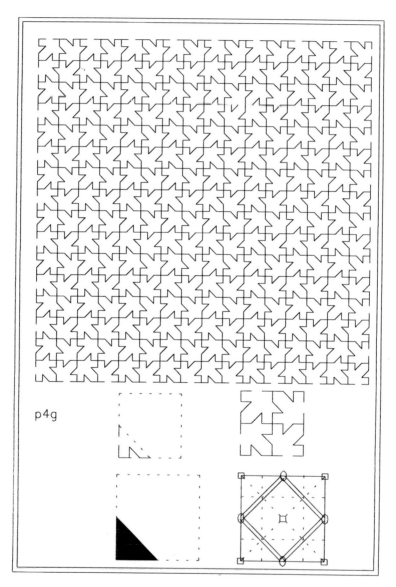

Figure 4.17. From the Great Mosque, Cordoba, Spain. The initial structure of the mosque was constructed in 785 A.D., but it was successively modified and enlarged over the next 3 centuries. The pattern is said to be an abstraction of the leaves of the fig tree and is one of the earliest in Islamic art. Photo 4.12.

4.2. Examples of the Seventeen Types of Patterns from Islamic Art 103

Figure 4.18: From the Alhambra, Granada, Spain; 14th century. Photo 4.13.

104   Chapter 4. Classification, Identification and Construction of . . .

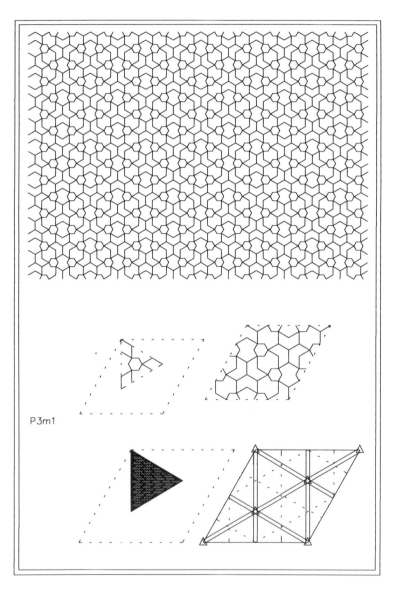

Figure 4.19: From Humayun's tomb Delhi, India; 16th century. Photo 4.14. (after Lalvani [24]).

4.2. Examples of the Seventeen Types of Patterns from Islamic Art   105

Figure 4.20: From the Tomb Towers Kharraqan, Iran; 11th century. Photo 4.15.

Figure 4.21: From the Great Mosque (La Giralda), Seville, Spain; late 12th century. Photo 4.16.

4.2. Examples of the Seventeen Types of Patterns from Islamic Art   107

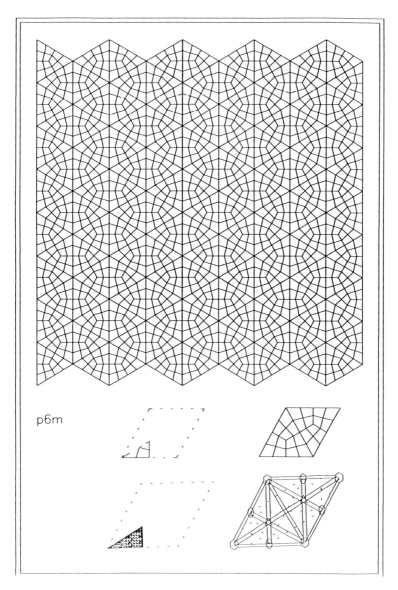

Figure 4.22: From the Top Kapi Sarayi Palace, Istanbul, Turkey; The Palace was the seat of the Ottoman rulers from the 15th to the 19th century. Photo 4.17.

a p6m type pattern which has a high degree of symmetry but is not at all exciting. Whereas the p1 type pattern in Fig. 4.5b, with nothing other than the translation symmetries, is obviously far more appealing.

Finally, it needs to be said that there are only six types of patterns that occur frequently in Islamic art, p6m and p4m being by far the most popular; see Fig. 5.1. However, all the seventeen types can be found, although it is difficult to find the types pg and pgg that have any real Islamic flavour. In our studies, we have not come across an example of an Islamic pg pattern in ground symmetry. In Fig. 4.9, which shows a pg type pattern, the interlacing, which destroys mirror symmetry, has to be taken into account.

4.3 IDENTIFYING THE SEVENTEEN PATTERN TYPES

The secret of Sherlock Holmes' success was his passion for acute observation. He frequently admonished his companion Dr. Watson with — *You see, but you do not observe, my dear Watson.*

We may not be able to match Holmes' zeal for acute observation, but we can certainly learn to identify the seventeen types of two-dimensional periodic patterns. The reader who tries this over some reasonable time span will find it a very worthwhile and enjoyable activity. Certainly, it is to be recommended to all students of science and particularly mathematics. Such patterns abound in the environment.

In learning to identify such patterns, perhaps the most important skill to develop is the observation of the underlying structure of the net on which a pattern has been built up. The square, the hexagonal, and the centered nets can usually be seen without requiring any careful observation. It is then often easy to settle the rest.

In going about it systematically, one can proceed step by step by discovering answers to the following questions:

1. Is there rotational symmetry about some points and, if so, what is the smallest angle through which the pattern can be rotated to make it coincide with itself?

2. Are there any mirror reflection lines?

3. Are there mirror reflection lines in more than one direction?

4. Are there any glide reflection lines? Do the glide lines coincide with mirror lines? Do centers of rotation lie on mirror lines?

Answers to some or all of these questions will clinch the issue. This method is explained in flow chart form in Figs. 4.23a–e, where we have also shown the basic structure of each type of pattern. The flow charts are based

4.3. Identifying the Seventeen Pattern Types

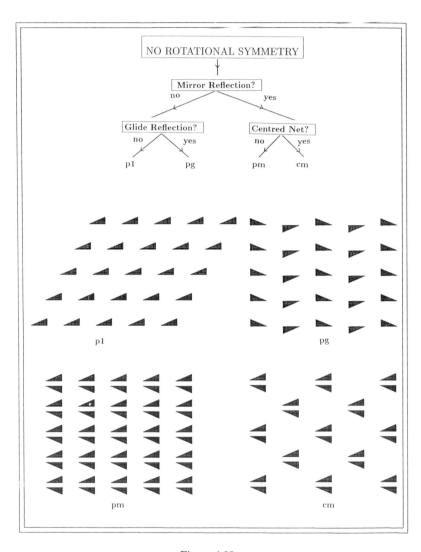

Figure 4.23a:

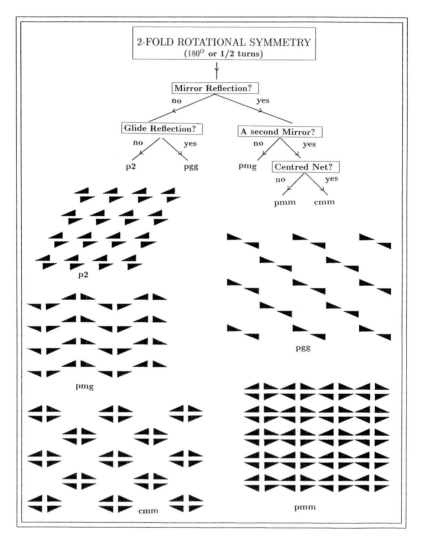

Figure 4.23b:

4.3. Identifying the Seventeen Pattern Types

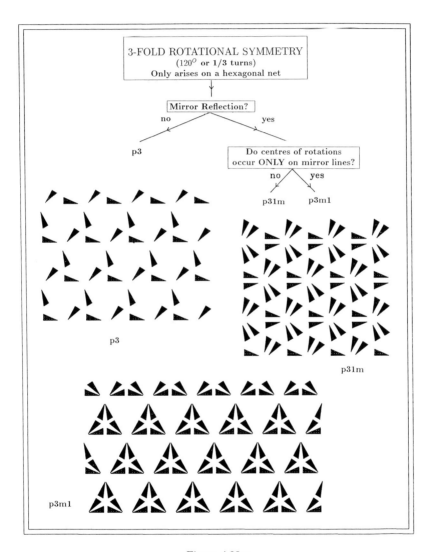

Figure 4.23c:

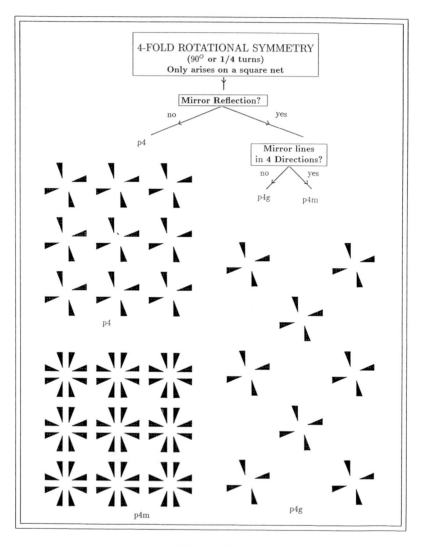

Figure 4.23d:

4.3. *Identifying the Seventeen Pattern Types* 113

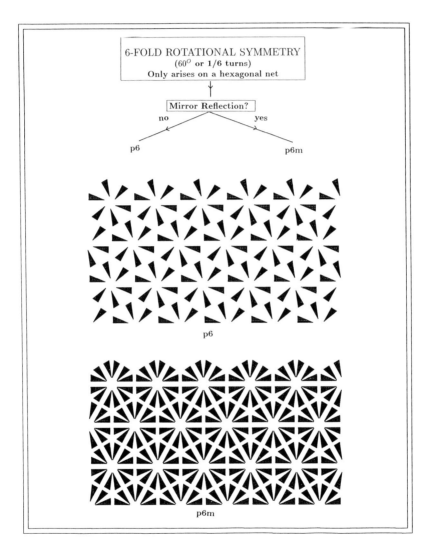

Figure 4.23e:

on a paper by Crowe [10] and the structure diagrams have been adapted from Shubnikov and Koptsik [44].

4.4 TILE-BASED ALGORITHMS FOR THE SEVENTEEN PATTERN TYPES

In this section, we shall give simple algorithms for the construction of each of the seventeen types of two-dimensional repeat patterns. The algorithms are particularly suited for use in interactive computer graphics, requiring only the facilities to rotate, reflect, and copy shapes. However, they can also be used to advantage when working with other media. A major virtue of these algorithms is that they show clearly the spatial constraints behind the various types of patterns. The algorithms are given in Figs. 4.24–4.40. The notation used is explained below. The reader will find other published algorithms in the references [1,2,25].

4.4.1 The Notation

The meanings of the symbols used in the algorithms are explained below and illustrated with an example that follows.

T
The symbol **T** denotes a tile in which a motif is to be placed. **T** will be called a **template** tile and the motif will serve as our template motif. It is assumed that the template motif has no symmetries, apart from the trivial one, i.e., a rotation by 360 degrees.

T_H and T_V
In the special cases when **T** has a horizontal or a vertical edge, the symbols T_H and T_V denote the tiles obtained by reflecting **T** in a horizontal and a vertical edge, respectively.

T_{AB}
In general, the subscripted symbol T_{AB} is used to denote a tile obtained by reflecting T in the line AB.

T_Π
The symbol T_Π means the tile obtained by turning **T** upside down.

4.4. Tile-Based Algorithms for the Seventeen Pattern Types

$\mathbf{T}_\theta^\mathbf{P}$

The symbol $\mathbf{T}_\theta^\mathbf{P}$ is used to denote the tile obtained by rotating $\mathbf{T}$ about the point $\mathbf{P}$ through θ degrees.

+

Finally, the symbol + has been used to denote the action of gluing two tiles together. Thus by $\mathbf{T} + \mathbf{T_H}$ we mean the composite tile obtained by gluing together the tiles $\mathbf{T}$ and $\mathbf{T_H}$.

The meanings of the subscripts and the superscripts, when used with any other tile, are the same as described above for the tile $\mathbf{T}$.

4.4.2 The Procedure

The procedure for producing a pattern type from the Figs. 4.24–4.40 is as follows:

1. Start by choosing the correctly shaped template tile shown and by placing a template motif in it.

2. Construct the **unit tile U** by gluing to $\mathbf{T}$ the transformed versions of $\mathbf{T}$ as shown.

3. Construct as much of the pattern as desired by simply tiling with $\mathbf{U}$ using translations and copying.

4.4.3 An Example

We see from the above that all we need to know is how to construct the correct unit tile $\mathbf{U}$ for each type of pattern. This is what is prescribed by the algorithms depicted in the Figs. 4.24–4.40. Consider, for example, the algorithm for constructing a p4g type pattern given in Fig. 4.35.

Pattern Type: p4g

$\mathbf{T} = $ **Any right angled isosceles triangle**

$\mathbf{T1} = \mathbf{T} + \mathbf{T_{BC}}$

$\mathbf{U} = \mathbf{T1} + \mathbf{T1}_{90}^O + \mathbf{T1}_{180}^O + \mathbf{T1}_{270}^O$

The above is to be interpreted as follows:

1. Choose a template tile **T**, in the shape of any right angled isosceles triangle. Place a motif in it and position **T**, as shown.

2. Obtain a tile **T1** by gluing together **T** and the tile obtained by reflecting T in the edge BC, as shown.

3. Obtain the unit tile **U** by gluing to **T1** the tiles obtained by rotating **T1** about the point O through 90, 180, and 270 degrees.

4. Tile with **U** using translations.

Note: (i) In the Example above the edges of the tiles **T**, **T1**, and **U** serve only to force the structure of the underlying net and the existence of the required symmetries. They need not be drawn in the pattern and, indeed, in most cases, they will be ignored. Another related feature is that there is no requirement for the template motif to lie inside **T**. The same type of pattern will be produced if the motif extends beyond the boundaries of **T**. Indeed, the template motif may lie entirely outside the template tile **T**. An example is shown in the case p1 in the Fig. 4.24.

(ii) Although a parallelogram shaped unit cell always exists, in the case of a hexagonal net it is sometimes more convenient to form the unit tile **U** in the shape of a hexagon. For some of the patterns which occur on the hexagonal net, we have constructed **U** as a hexagon and for some others which occur on this same net, we have given two forms for **U**, one a rhombus and the other, **U2**, a hexagon.

4.5 CONCLUDING REMARKS

In this chapter we explained how two-dimensional periodic patterns are classified into seventeen different types and how the various types can be identified. Examples of all the seventeen types from Islamic architectural decorations were presented. We also gave simple algorithms which show clearly the spatial constraints and the structure of each type and which may be utilized to generate them using computer graphics.

The next chapter will present a large collection of Islamic patterns on which all the skills of this chapter may be practiced at length.

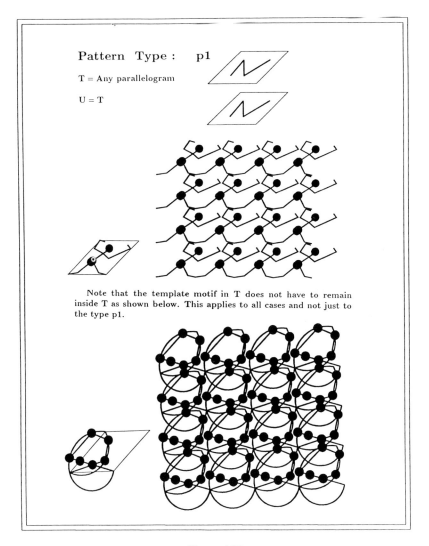

Figure 4.24:

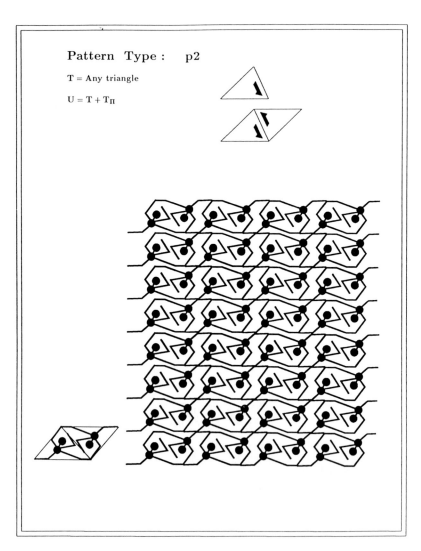

Figure 4.25:

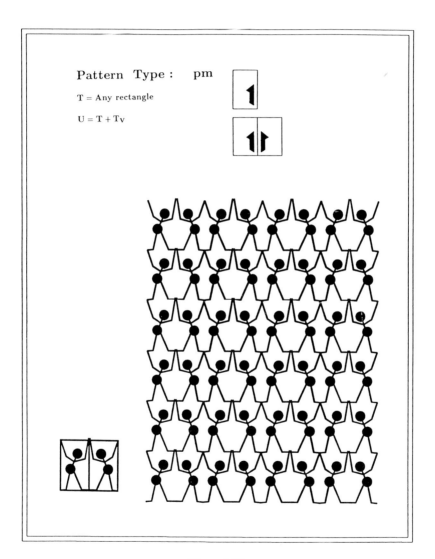

Figure 4.26:

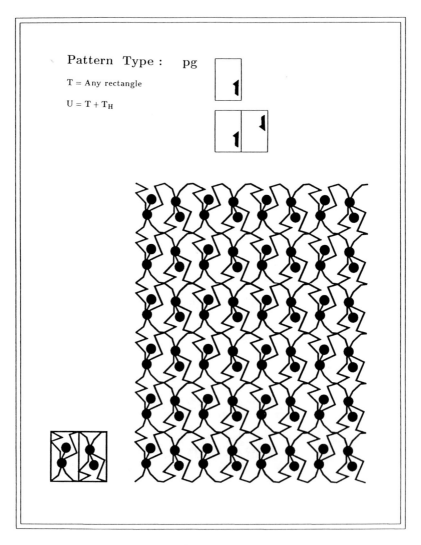

Figure 4.27:

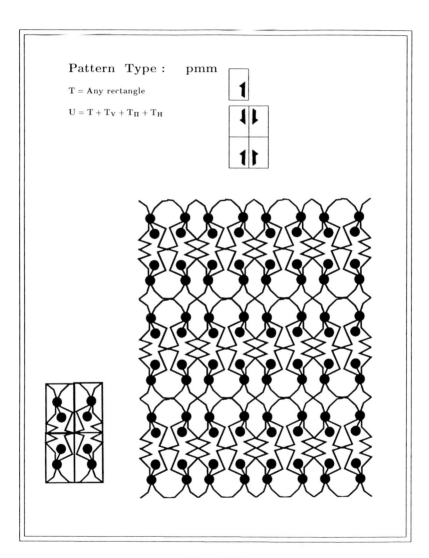

Figure 4.28:

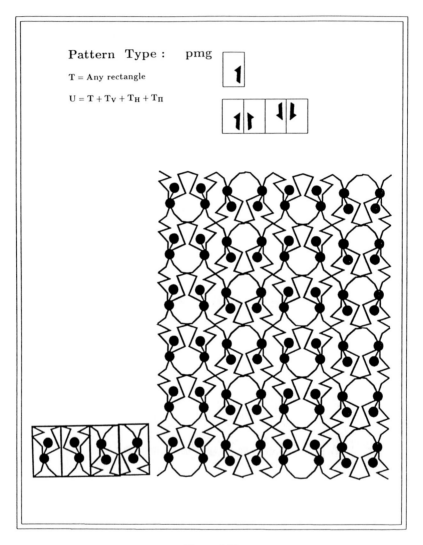

Figure 4.29:

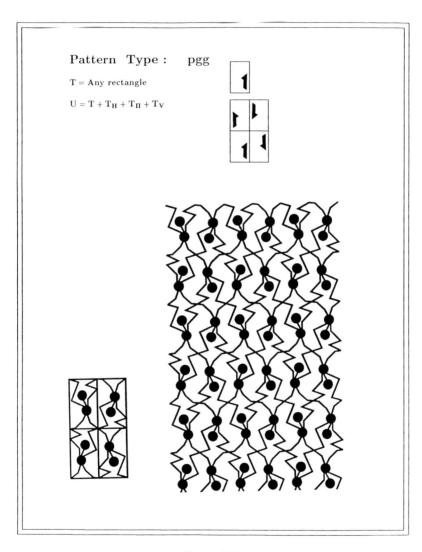

Figure 4.30:

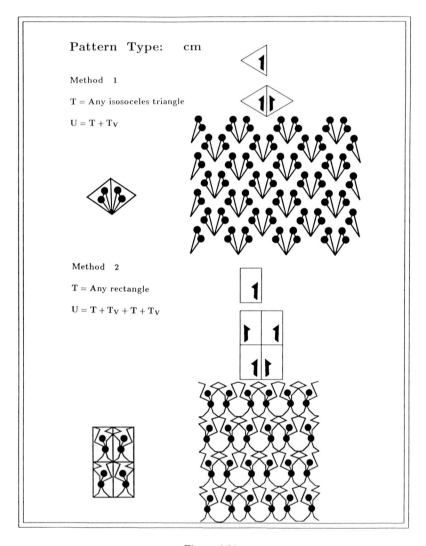

Figure 4.31:

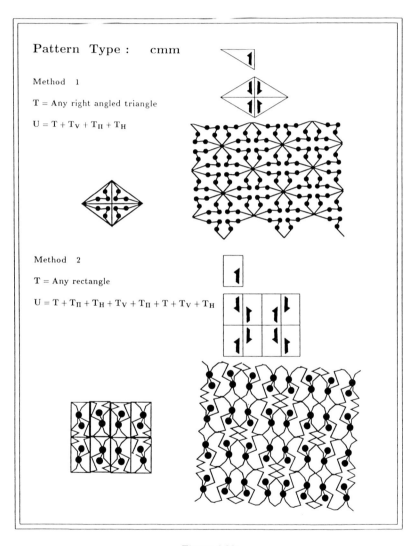

Figure 4.32:

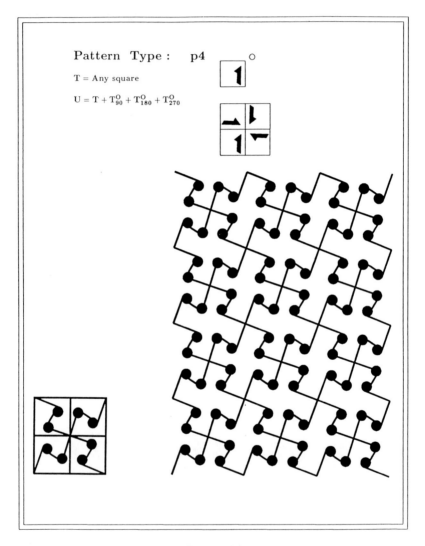

Figure 4.33:

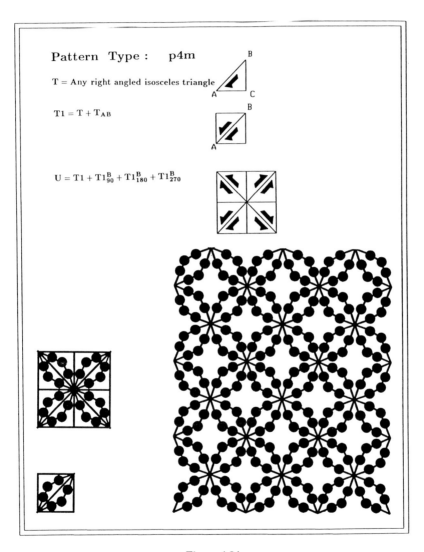

Figure 4.34:

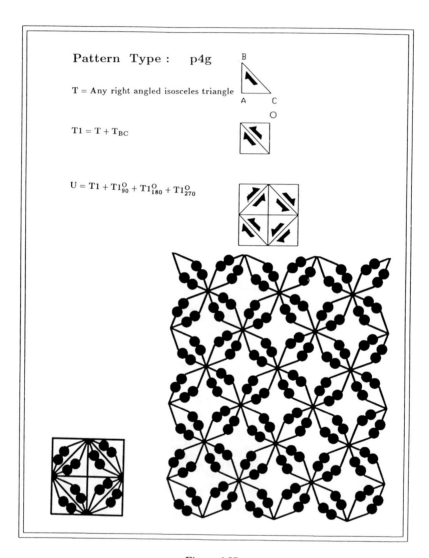

Figure 4.35:

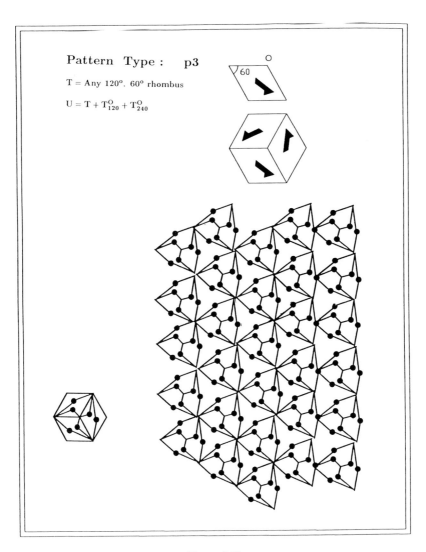

Figure 4.36:

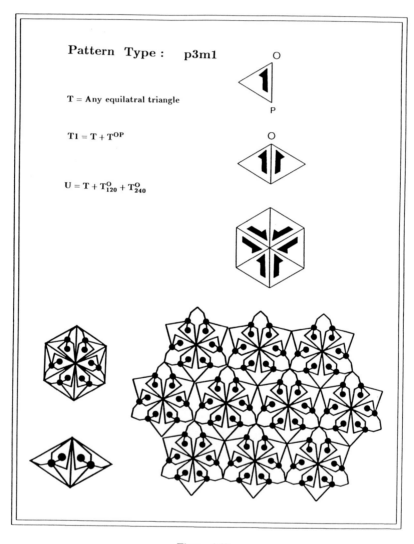

Figure 4.37:

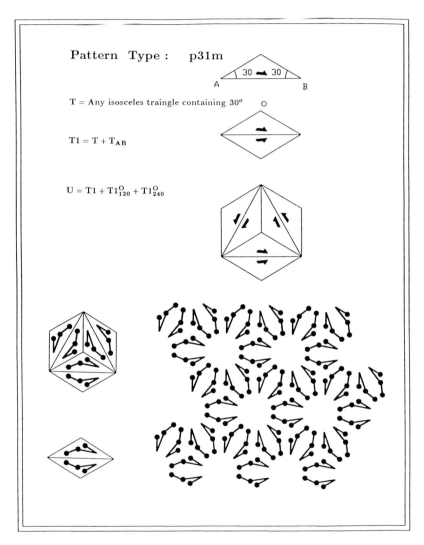

Figure 4.38:

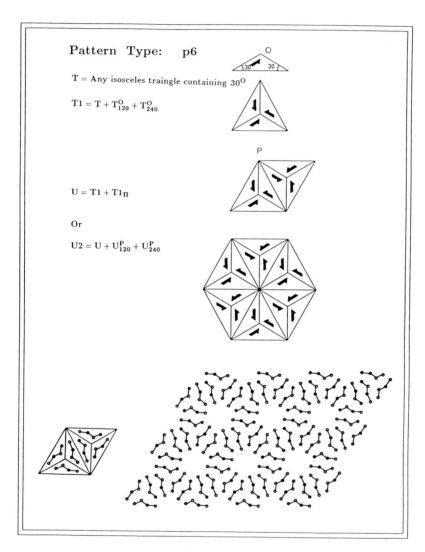

Figure 4.39:

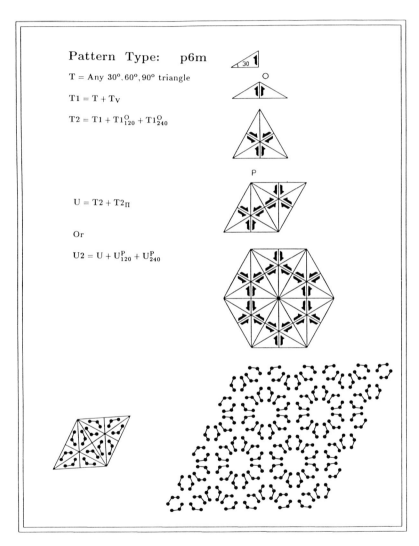

Figure 4.40:

Chapter 5

ISLAMIC PATTERNS AND THEIR SYMMETRIES

> *And as in the case of Islamic art in the past, science must come to the service of the arts, whether we are talking today of Islamic art, of Western art or of art generally, today more than ever before, for otherwise I cannot imagine how the arts can move into the twenty-first century.*
>
> *Chorbachi*

In this final chapter we catalogue and present, according to symmetry groups, a substantial collection (250) of Islamic patterns. We also show the template motif and the unit cell of each pattern. The catalogue and the analysis arise from our computer graphic studies of more than 350 Islamic geometrical patterns over the past several years.

Our work was begun by analyzing the symmetries of the patterns given in the books by Bourgoin [6], Critchlow [9], El-Said & Parman [13], and Wade [50]. We again acknowledge our major debt to these authors.

Once the selected patterns had been analyzed and placed into separate groups, we reconstructed them using traditional geometrical methods. Instead of using rulers and compasses, however, we utilized a CAD package. The reconstructions were used to explore the geometrical structures of the patterns and also to extract numerical information for the description of the template motifs. After this a computer program was written to construct the patterns automatically.

Subsequent to our initial studies based on the publications of the authors mentioned above, we have examined numerous other works which are listed in the references and have also undertaken study trips to Spain, Morocco, Tunisia, Turkey, and Egypt. Several improved versions of computer algorithms have been constructed and implemented in graphics programs.

There are several theoretical as well as practical purposes in offering this collection. One purpose is to relate Islamic patterns to more advanced mathematics than has been done previously and to encourage the exploitation of their potential in mathematical education. The analysis of symmetry and the extraction of the template motif can serve as valuable exercises in mathematical education. They can be used to develop the "analytic eye" and the faculty for discovering unsuspected relationships, these being two of the key qualities of a mathematician.

In more general terms, the presentation is intended to encourage the bringing of science to art and art to science.

The patterns, as presented in the collection here, are in their bare essential form. They show only the basic geometrical structures. The template motif, as should be fully understood by now, contains the minimal geometrical information needed to generate the pattern. This, together with the knowledge of the symmetry group, prescribes how a pattern can be made most economically.

The imagination must come into play when it comes to producing art. Even the simplest of the patterns given here can be made to come to life in a variety of ways through the use of color, shading, interlacing, etc. Even a quite basic one can be modified to produce complex and interesting new ones. The possibilities for exercising the imagination are boundless. We have shown an example of color rendering in plate 13 of the first two patterns in the collection. Plate 14 shows some wallpaper patterns produced commercially and is intended to display the kind of effects that are possible.

Computer graphics and CAD may be utilized to produce the patterns on all kinds of materials — wood, plastic, glass, and so on. For such work numerical data will be needed. The numerical data associated with the patterns offered in this chapter is not included in the book. It is too cumbersome and will not be of interest to most of our readers. We are considering the possibility of publishing it separately.

5.0.1 Related Works

We have almost reached the end of the book. It may be useful before closing to offer the reader a brief history of related works on the subject

5.0.1. Related Works

and references to them.

It was in the early part of this century that Polya [40] and Speiser [45] instigated group theory based analysis of decorations and designs of various cultures. Speiser drew particular attention to Islamic art.

Not surprisingly, the Alhambra with its great profusion of patterns, was the first to attract attention. In 1944 Edith Muller [33] carried out an extensive symmetry analysis of the designs to be found in there. She discovered 11 different groups.

Following Muller, the Palace of Alhambra was scrutinized by several other mathematicians over the next three decades. Some of whom claimed, without offering proof, to have found all the 17 types. This led to the matter becoming obscure and controversial.

Then in the early 1980s, B. Grünbaum,[1] Z. Grünbaum of the University of Washington in the U.S.A, and G. C. Shephard of the University of East Anglia in England made a special trip to try to settle the issue. The results of their investigation appeared in print in 1986 [14]. They found 13 different groups; the four that they failed to discover were the types pg, p2, pgg, and p3m1. Very soon after, however, R. Pérez-Gómez [39] of the University of Granada and also J. Montesinos [32] of the University of Madrid found the four missing ones. The truth of the matter, to the satisfaction of interested mathematicians, is at last now settled. All the 17 types do exist in the Alhambra.

Apart from the patterns in the Alhambra, Islamic patterns from other places have also been analyzed previously by several authors. Harry Bixler, in a very interesting Ph.D. thesis submitted to New York University in 1980 [5], applied group-theoretic analysis to Islamic patterns. He carried out an extensive scrutiny of Islamic works of art in the Metropolitan Museum of Art in that city. Haresh Lalvani of the School of Architecture, Pratt Institute, Brooklyn, New York has made a comprehensive study of Islamic patterns from India [23]. He has also studied Islamic and other patterns using computer graphics. E. Makovicky and and M.Makovicky of the Institute of Mineralogy in Copenhagen are two other persons who have been very active [26] on symmetry analysis of Islamic patterns. They have analyzed patterns from a wide range of countries in the Islamic world.

We have gained much from the scholarship of all these authors and as recorded in the Acknowledgments, have also received help and support from several of them.

---

[1]B. Grünbaum and G. C. Shephard are the authors of the definitive treatise on tiling theory [15]. Their book contains many references to Islamic patterns.

5.0.2 Preferred Symmetry Types in the Islamic Culture

In chapter 2 we mentioned the recent discovery of Dorothy Washburn and Donald Crowe [51], that only certain symmetry types are preferred and recognized as being right in each culture. Figure 5.1 shows our findings on the preferred symmetry types in Islamic cultures based on our examination of more than 350 patterns. The types p6m, p4m, cmm, pmm, and p6 are the preferred symmetry types with p6m and p4m dominating. Some, such as pg and pgg, are very rare. There are several points that call for comment.

The collection offered in this chapter does not have the same distribution as that in Fig. 5.1. This is because we were obliged to limit the size of the book and decided to exclude patterns from our total collection which were either very similar or not very attractive to our eyes. One example each of the types pg and pgg has been included purely for completeness; the patterns are not particularly Islamic.

Finally, it should be remembered that our analysis in Fig. 5.1 refers to ground symmetry (see chapter 4). If we allow calligraphy, color, interlacing, and other decorations then of course the conclusion would be quite different. For example, in that case there would be no shortage of p1 type.

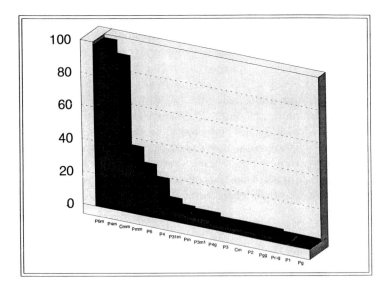

Figure 5.1: The distribution of various symmetry groups in islamic patterns.

5.1 CONCLUDING REMARKS

In this book we set out to show the universal relevance of the gifts of Islamic art, of pattern, symmetry, and unity of science and art. We hope that we have made some small contribution but that is the privilege of our reader to decide. We shall bring this book to an end by repeating some passages from the Preface.

Our greatest satisfaction would come if we were able to feel that we managed to encourage the regeneration of the Islamic art of pattern and symmetry through the use of computer graphics:

Pattern making in Islamic art reached its zenith in the mid 14th century resulting in the Alhambra and then, sadly, this profound and marvelous activity lost its vigor and spent itself. It is true that in some places, particularly Morocco, the ancient skills are alive and flourishing but no real major innovation can be observed. The mathematicians in the Islamic cultures of today have little appreciation of the value of their tradition. Computer graphics combined with modern mathematics offer all sorts of new and exciting possibilities.

To be a little futuristic, we suggest that it would soon be possible to enter virtual reality to fly through patterned spaces, wall after wall, column after column, dazzling with color and light. There, a seeker after unity in science and art would be able to submerge himself or herself in exquisite Alhambras of the mind to experience the art of pattern and symmetry that is yet to come.

On this flight of the imagination we bid the reader farewell — *Tamam Shud*.

> *Up, O ye lovers, and away! 'Tis time to leave the world for aye.*
> *Hark, loud and clear from heaven the drum of parting calls —*
> *let none delay!*
> *The cameleer hath risen amain, made ready all the camel-train,*
> *And quittance now desires to gain: why sleep ye, travellers,*
> *I pray?*
>
> R. A. Nicholson (Persian Poems)

p6m

142

p6m

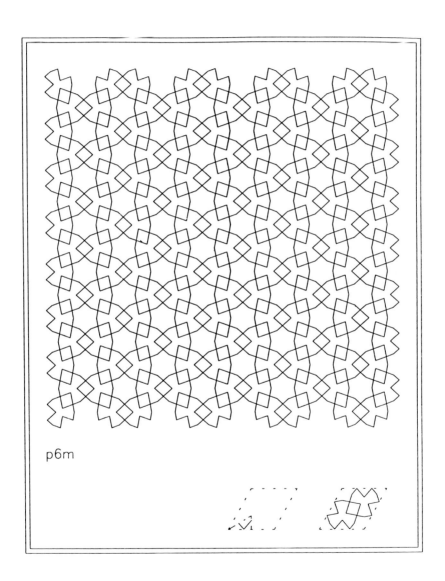

p6m

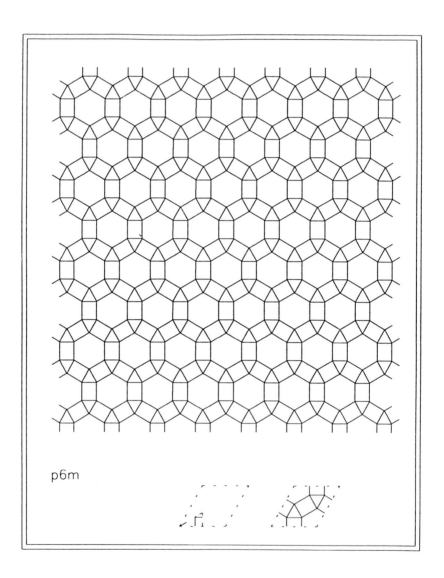

p6m

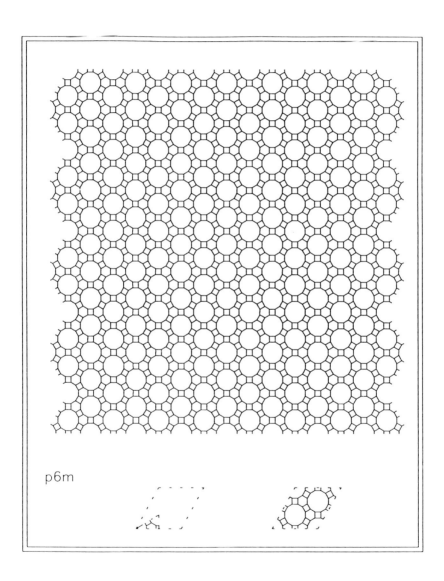

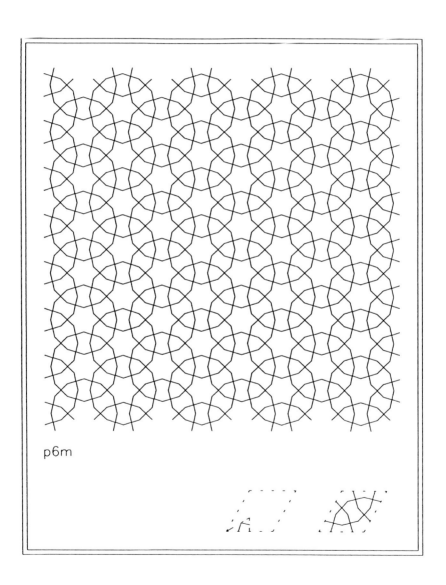

p6m

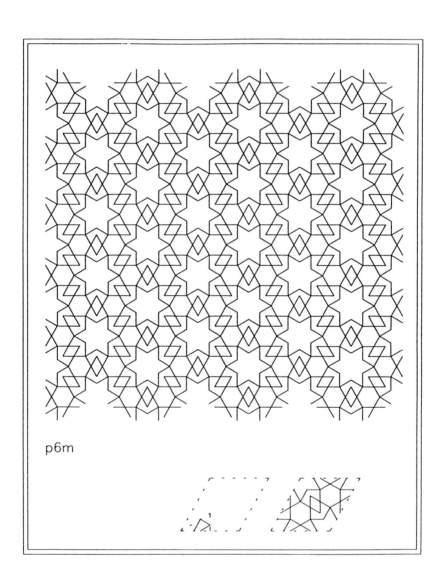

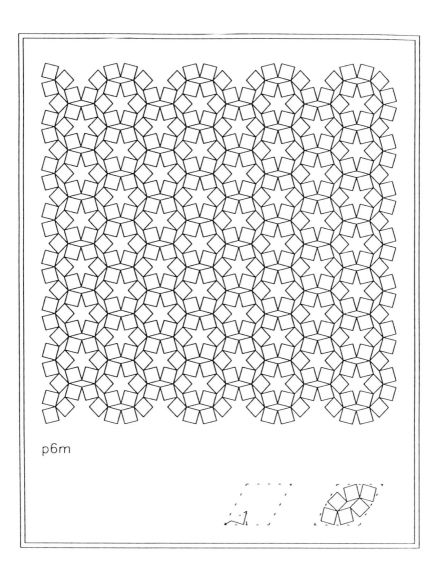

p6m

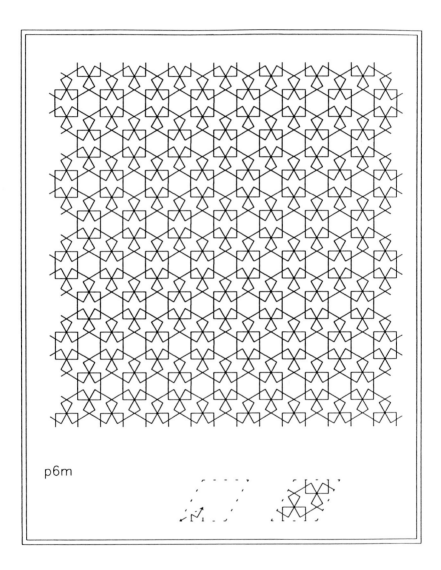

p6m

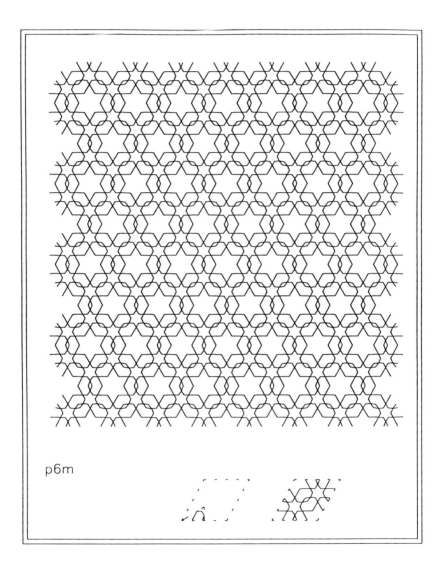

p6m

p6m

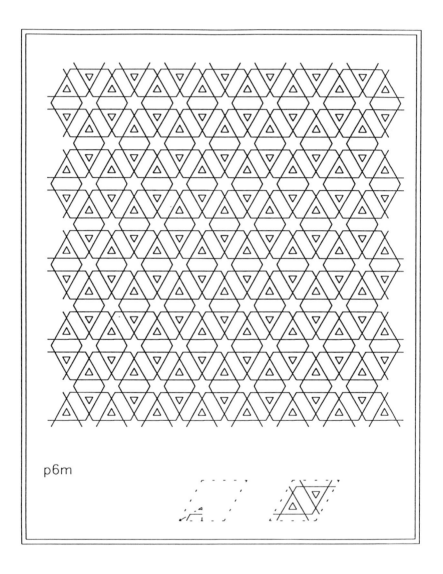

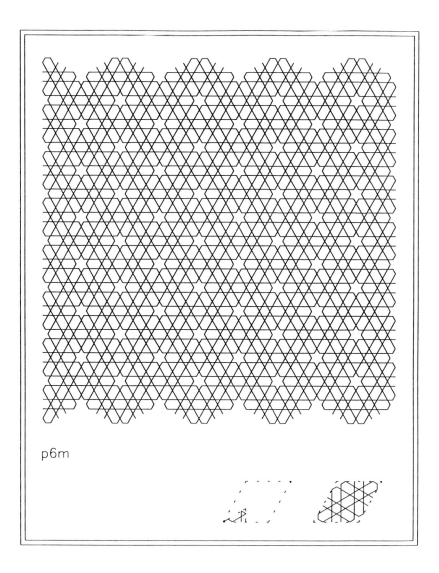

p6m

p6m

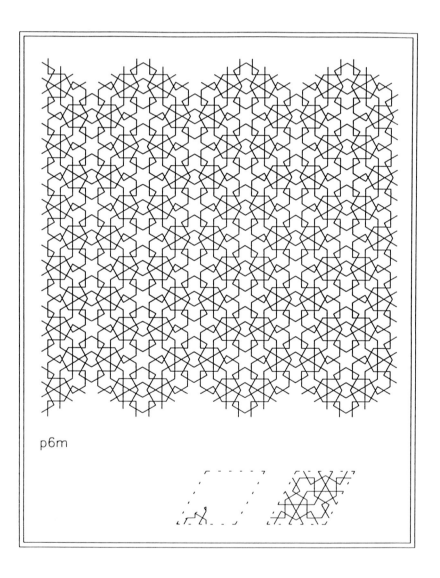

p6m

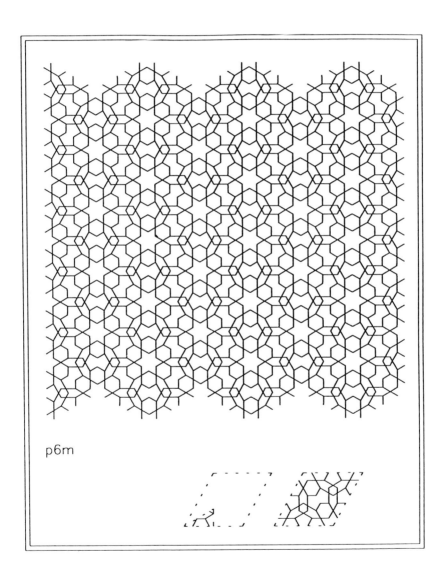

p6m

p6m

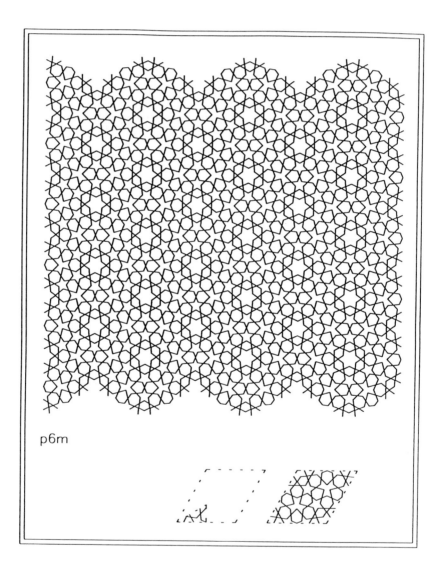

p6m

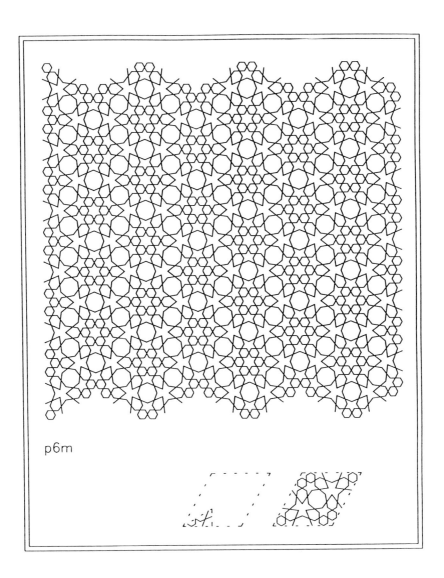

p6m

p6m

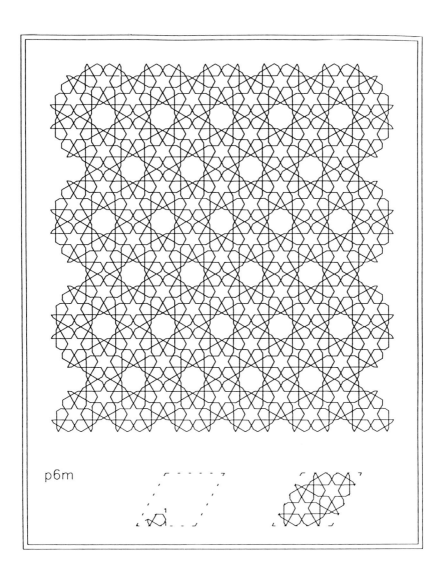

166

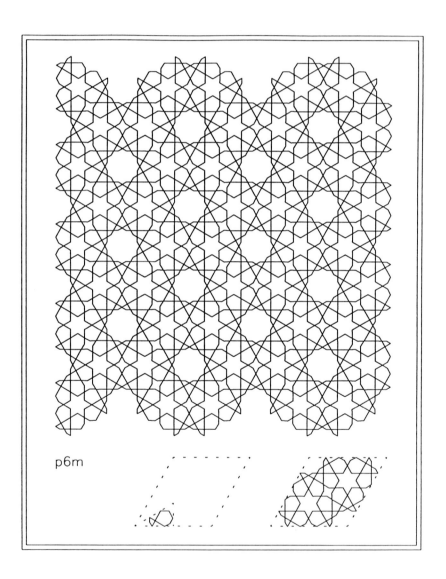

p6m

p6m

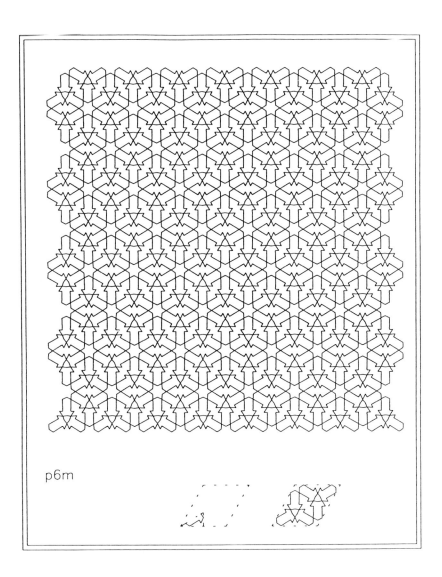

p6m

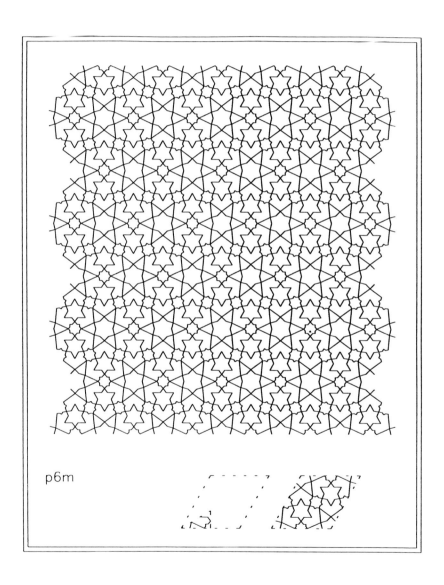

p6m

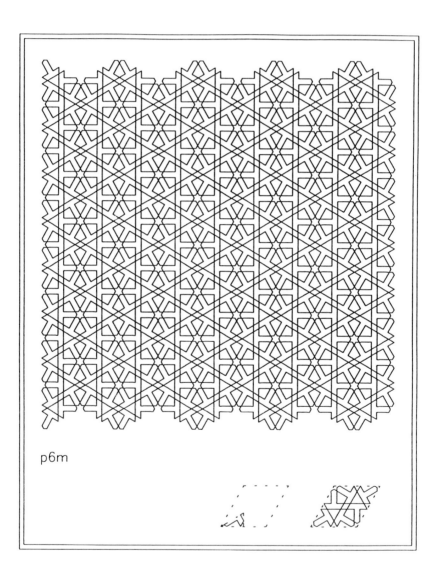

p6m

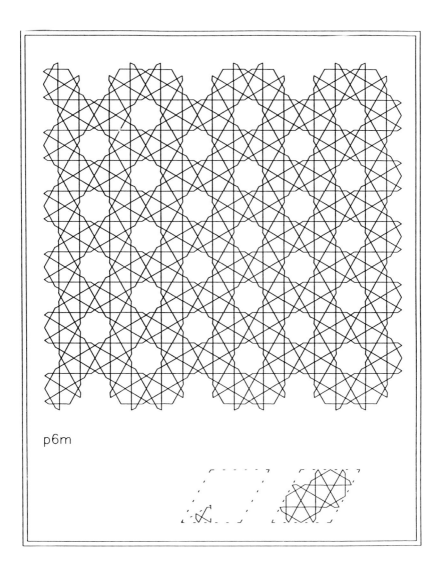

p6m

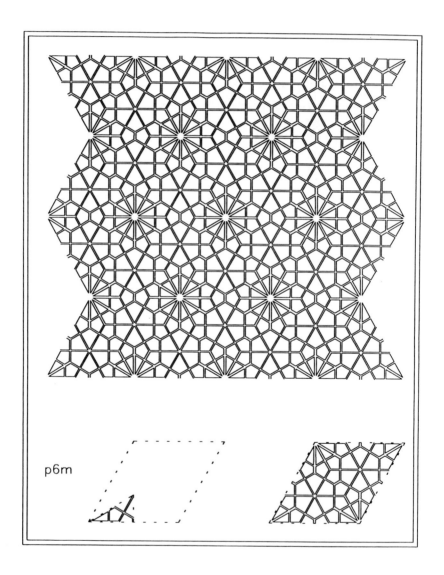

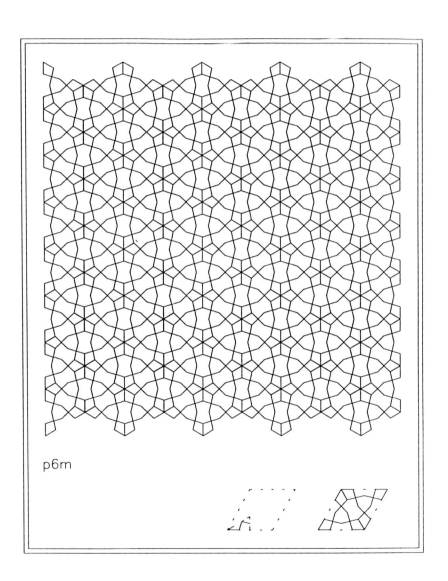

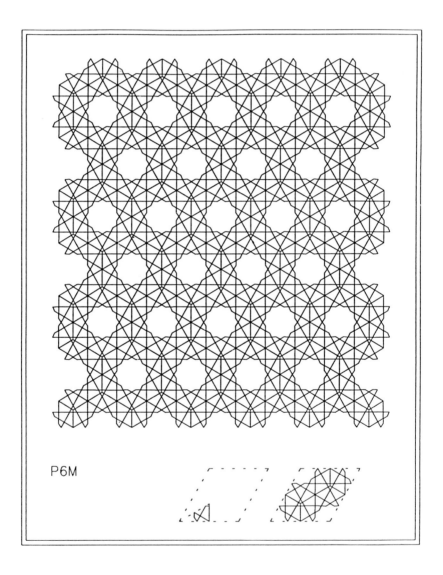

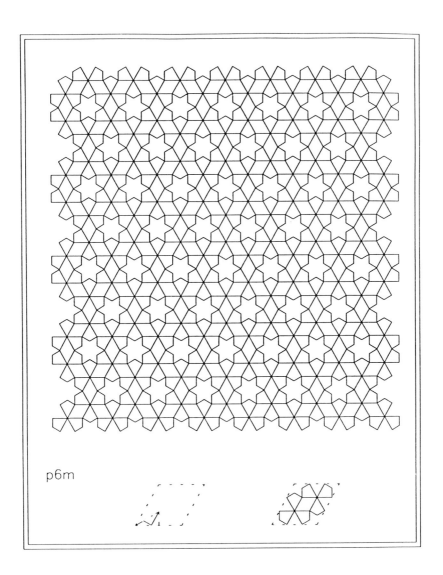

p6m

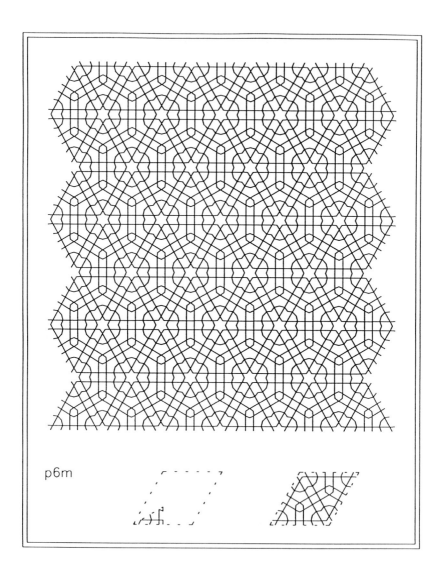

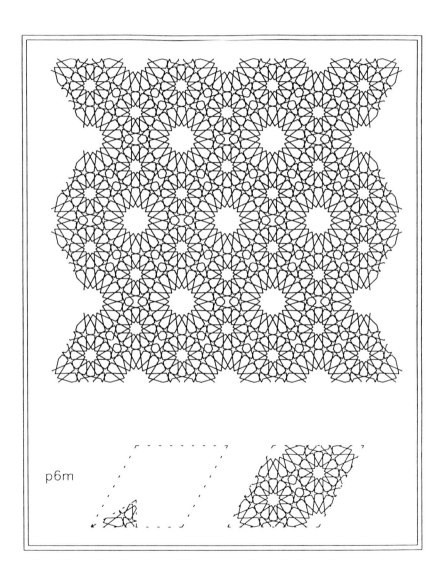

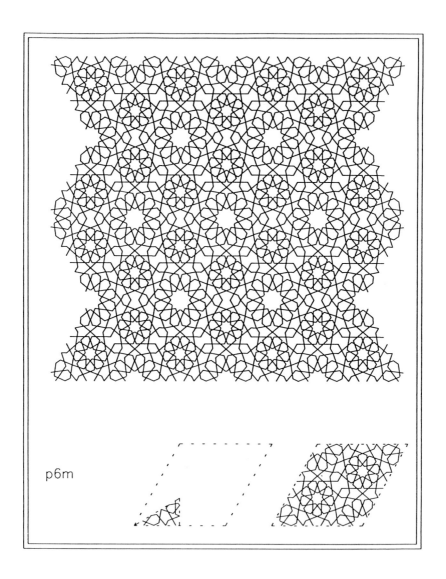

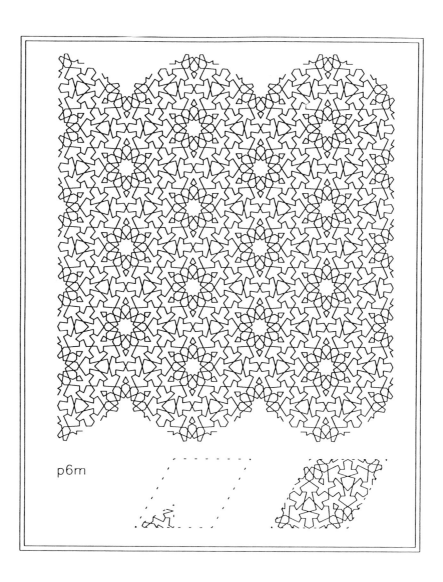

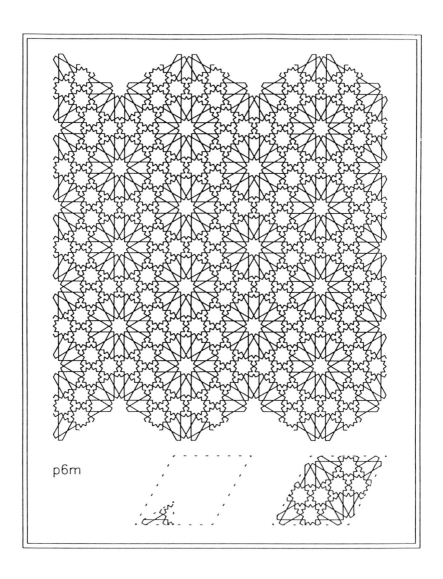

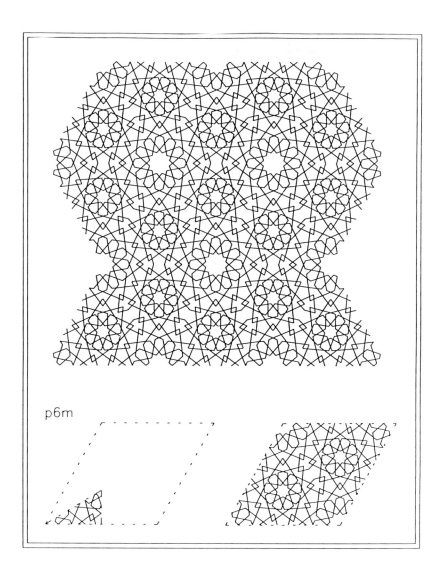

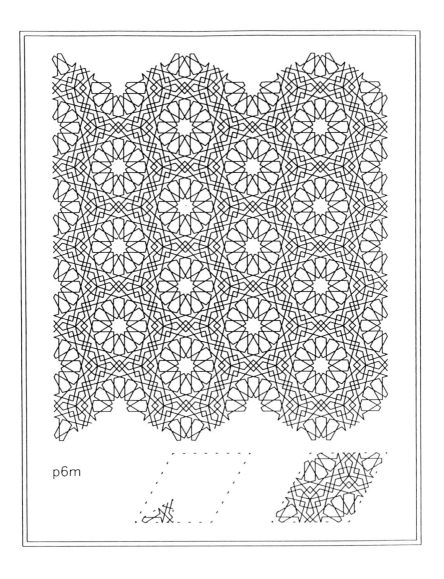

187

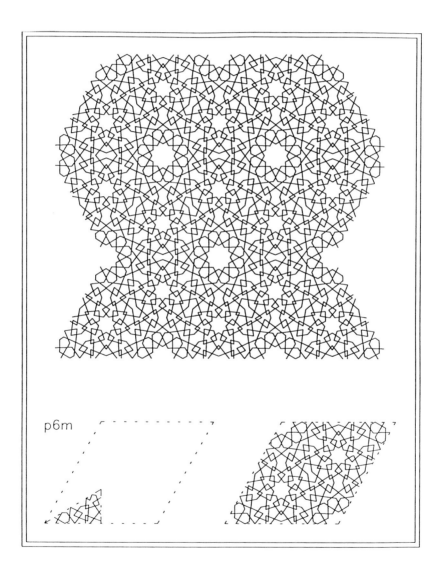

p6m

p6m

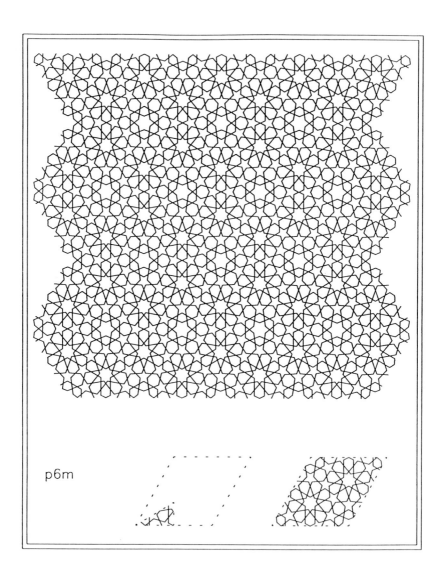

p6m

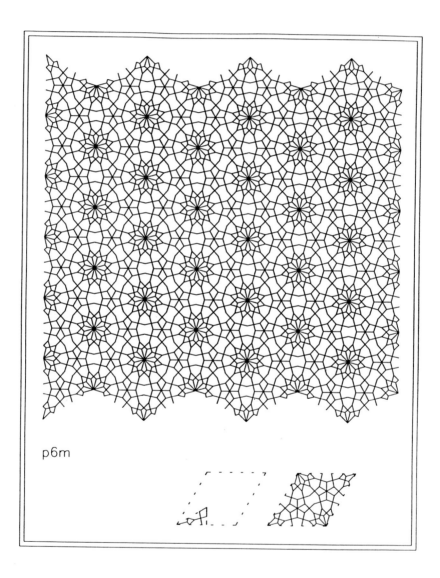

p6m

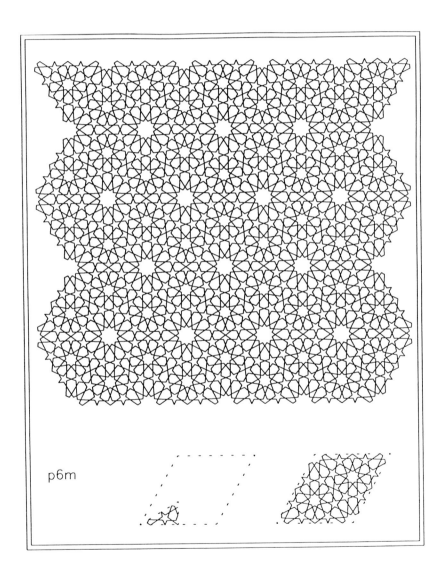

p6m

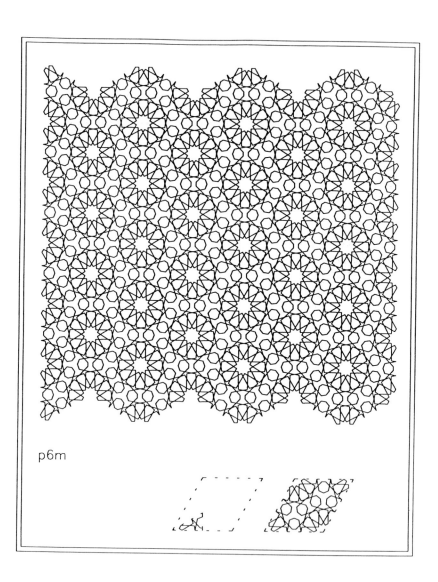

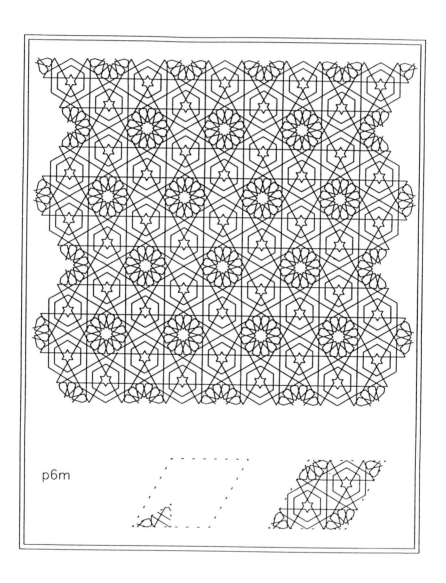

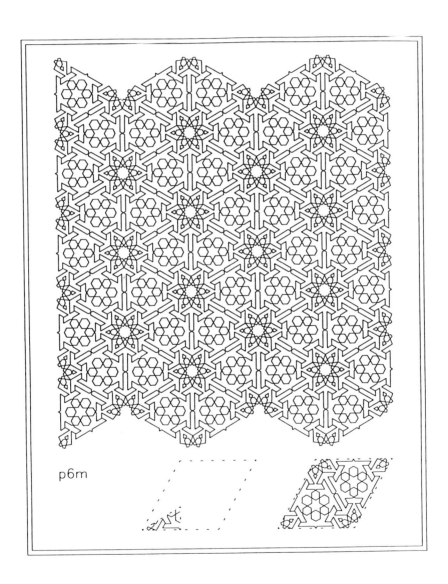

p6m

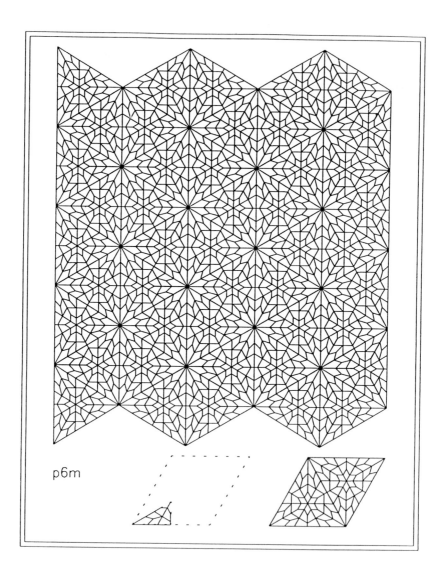

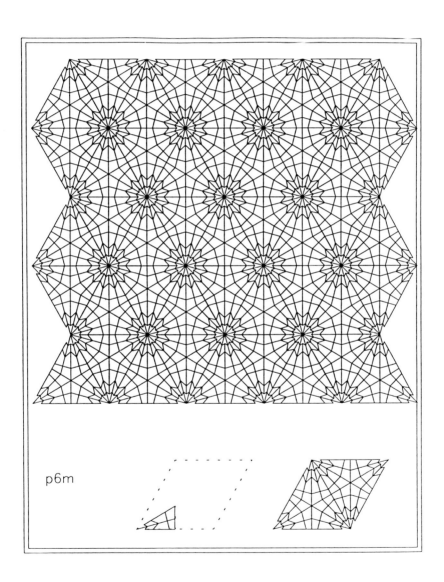

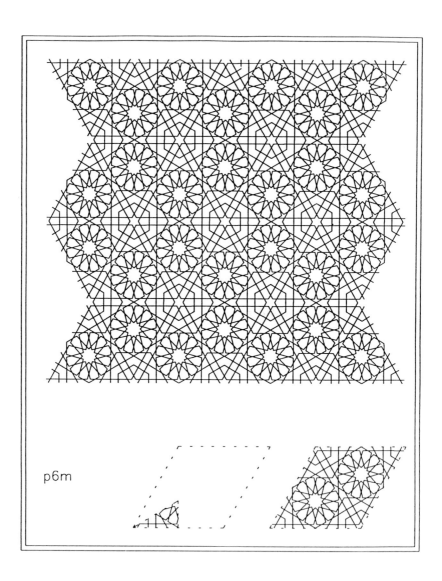

p6m

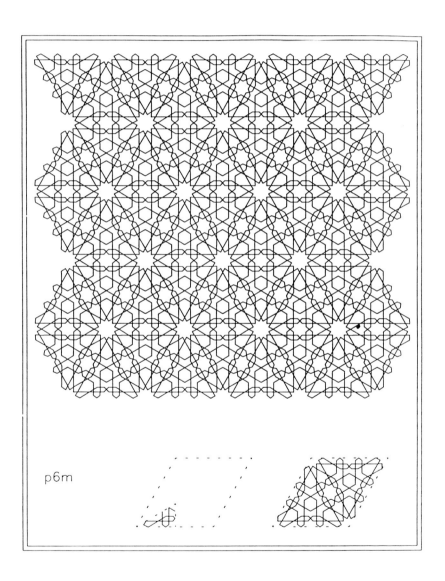

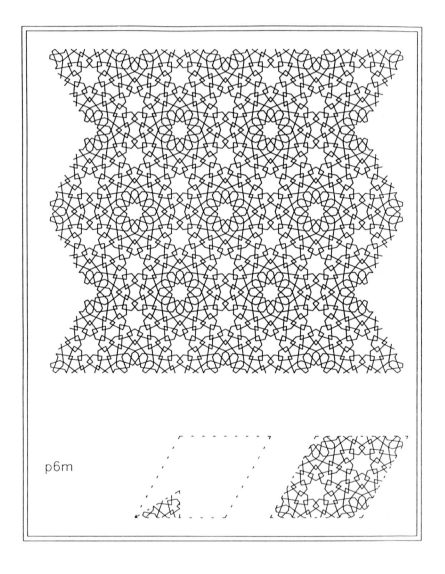

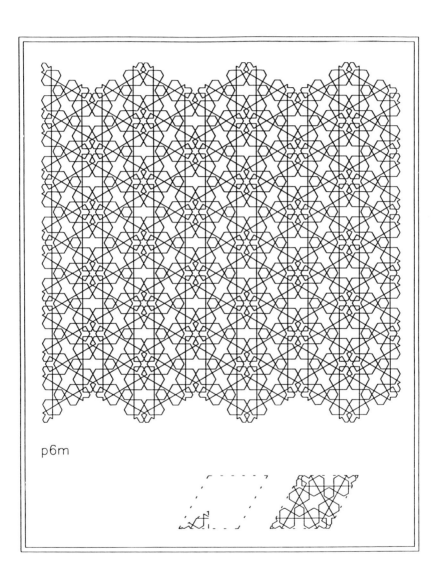

p6m

p6m

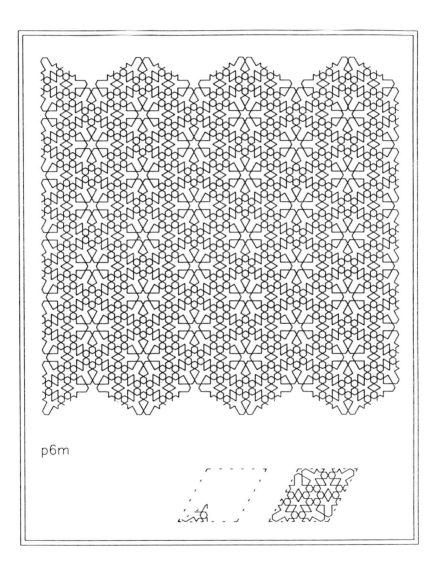

p6m

p6m

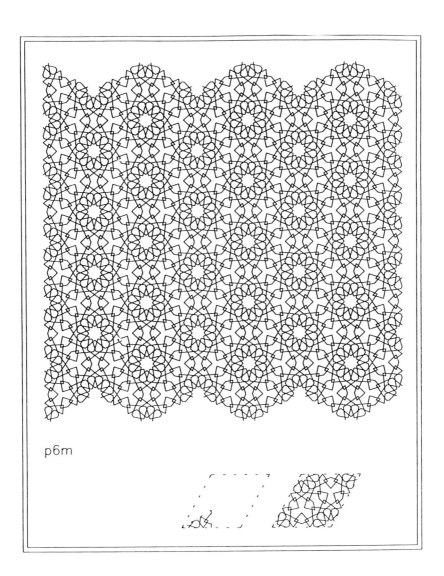

p6m

p6m

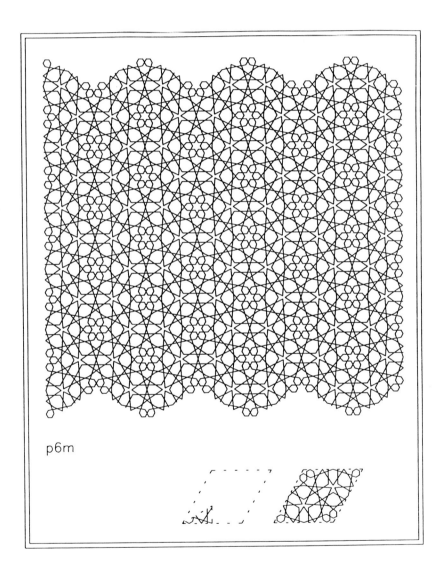

p6m

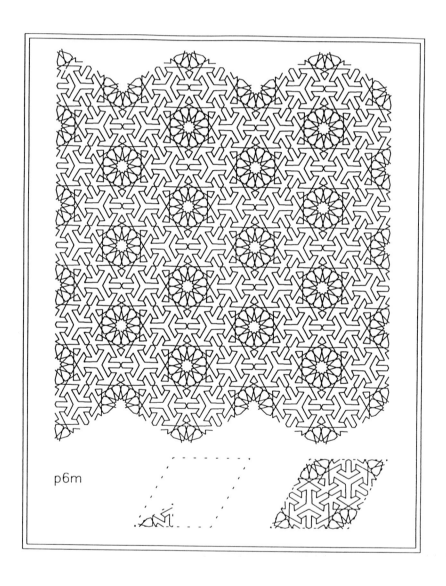

p6m

p6m

215

p6m

p6m

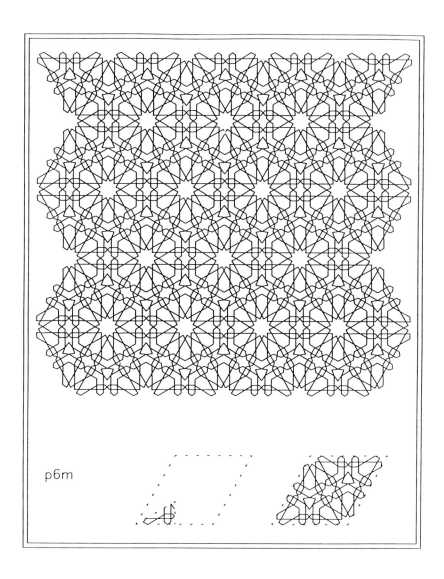

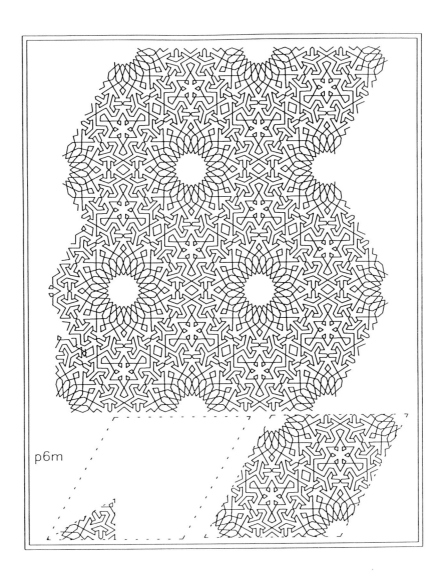

p6m

221

p4m

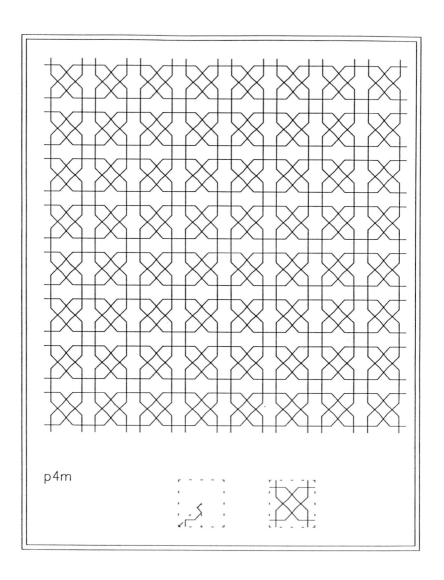

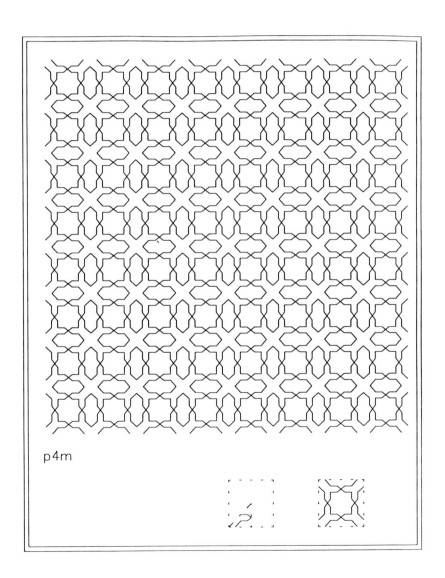

p4m

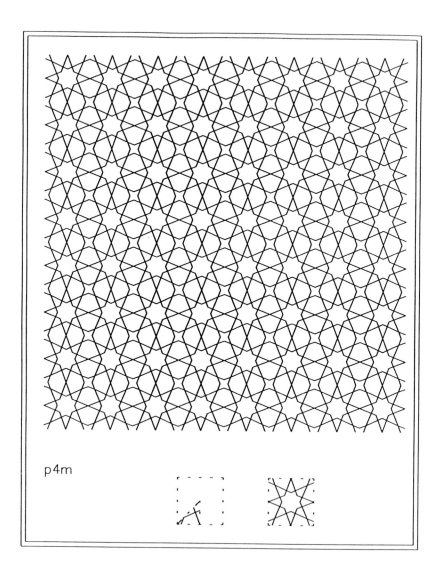

p4m

p4m

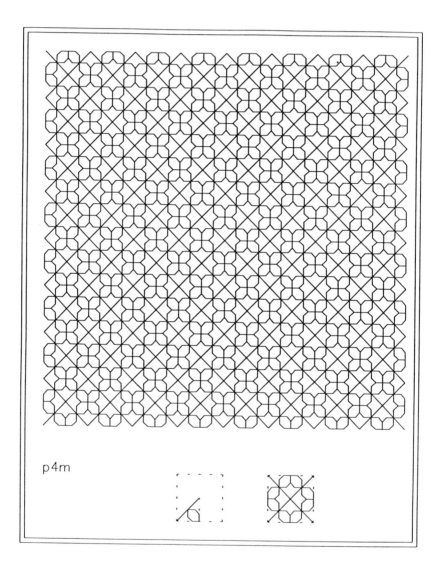

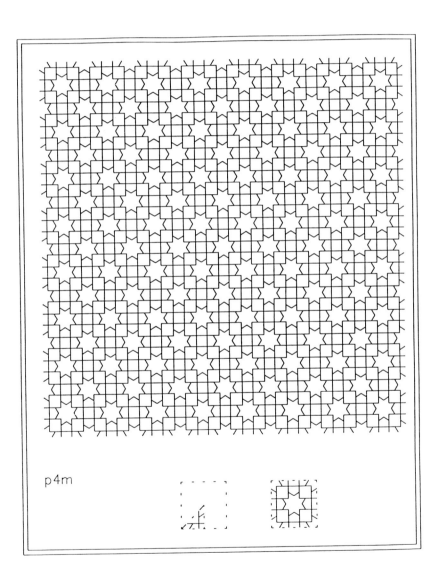

p4m

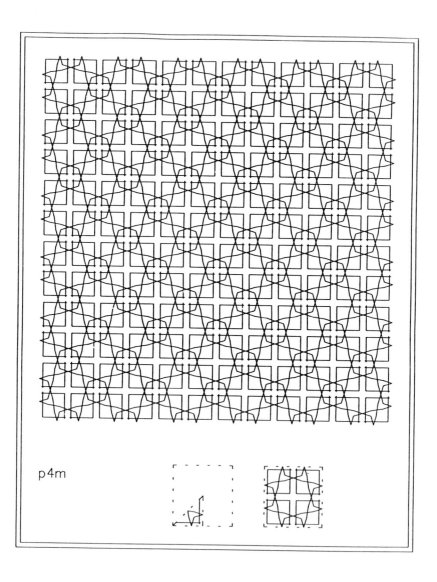

p4m

p4m

p4m

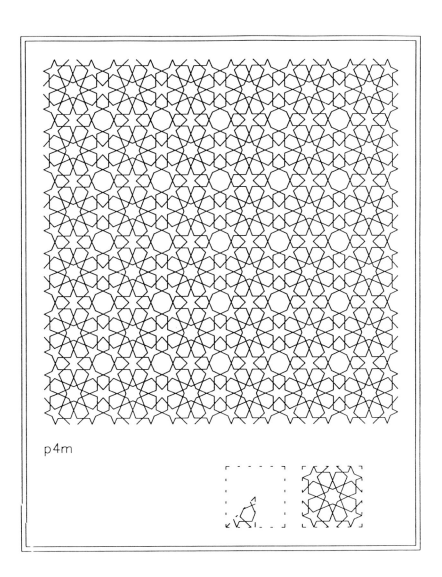

p4m

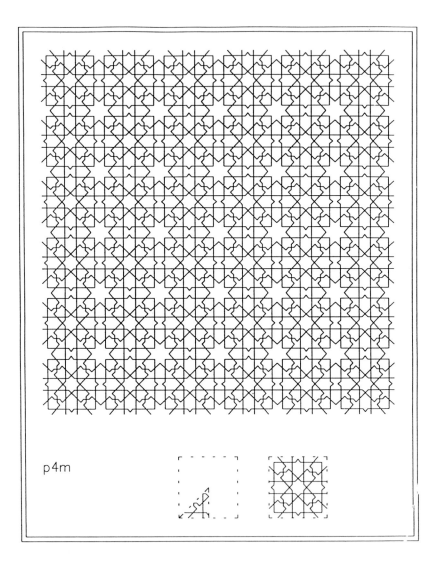

p4m

p4m

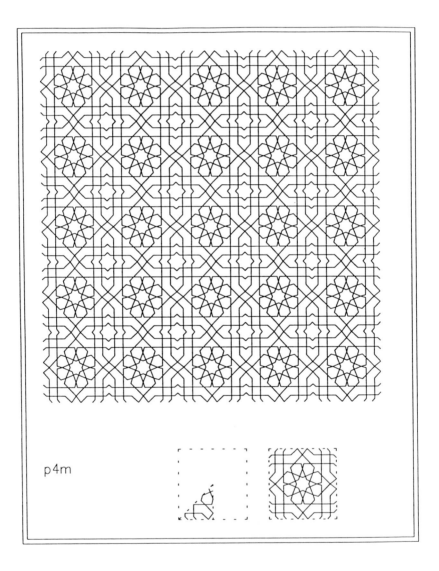

p4m

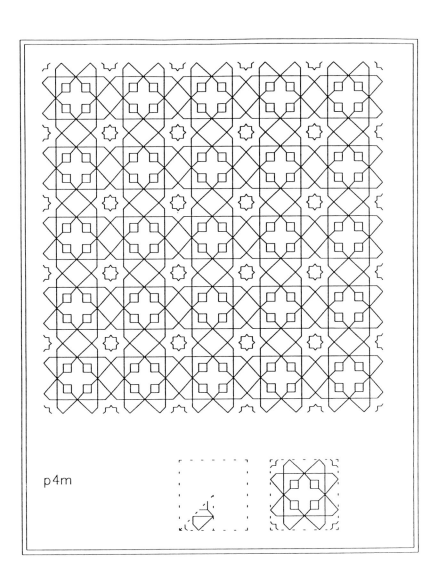

p4m

p4m

p4m

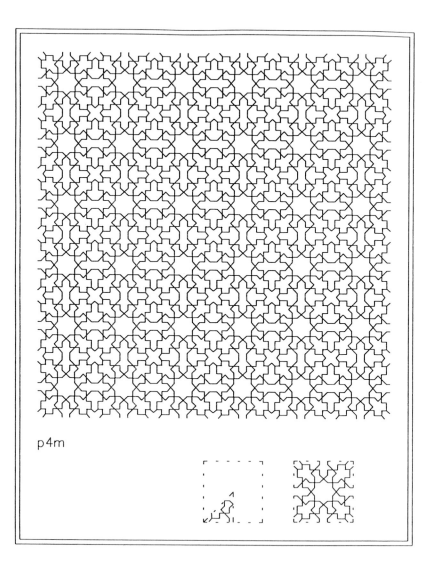

p4m

p4m

p4m

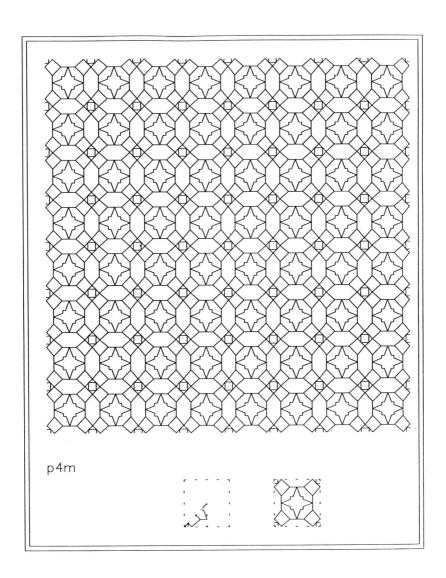

p4m

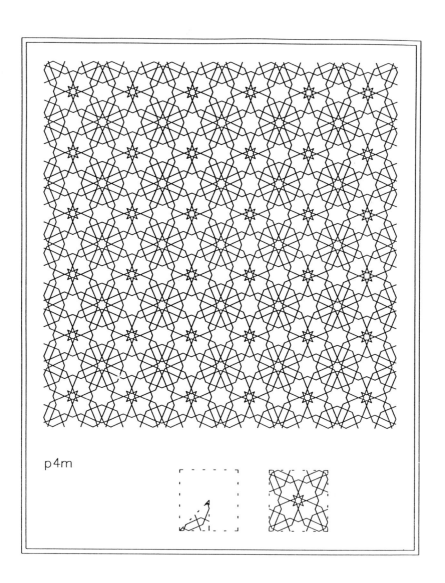

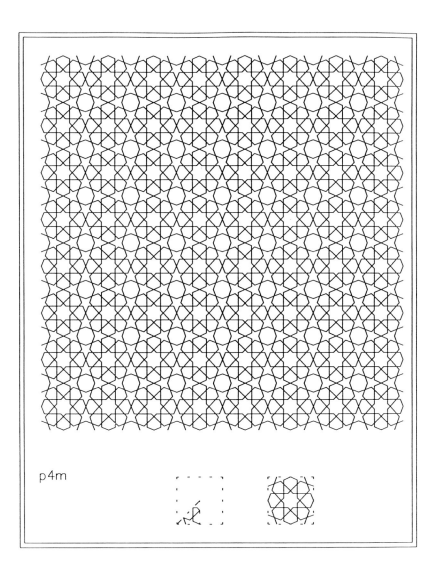

p4m

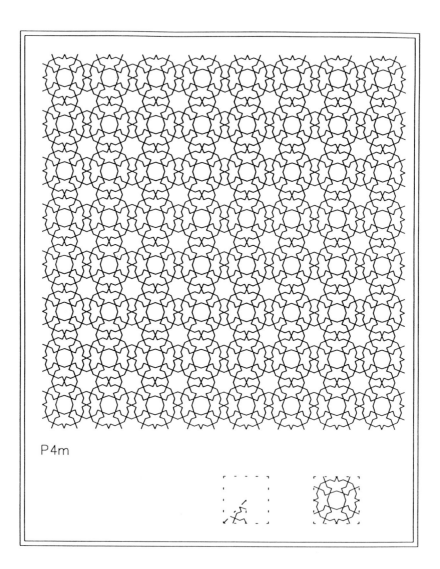

P4m

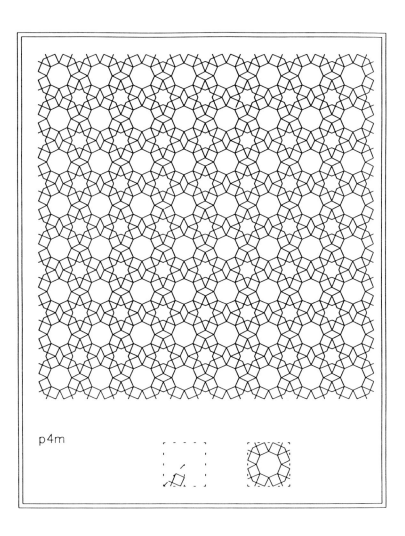

p4m

p4m

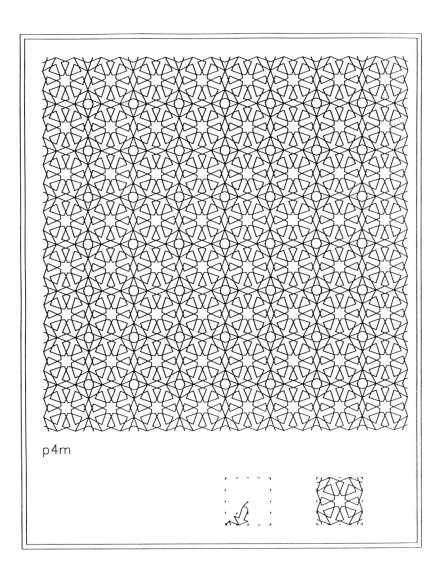

p4m

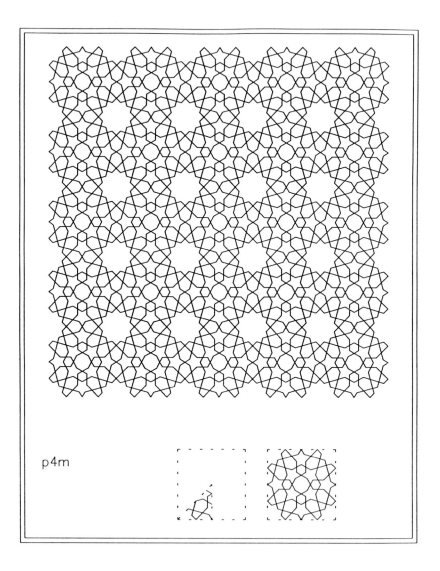

p4m

p4m

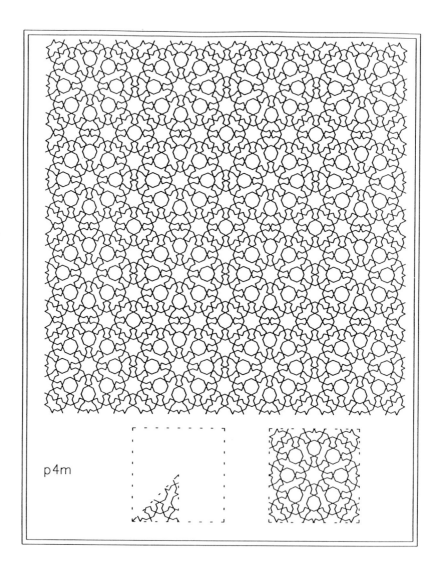

p4m

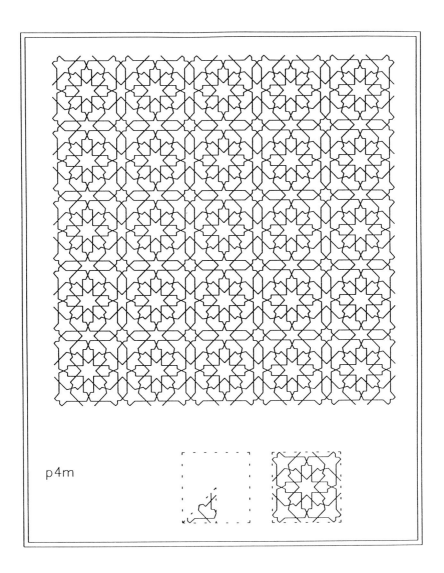

p4m

p4m

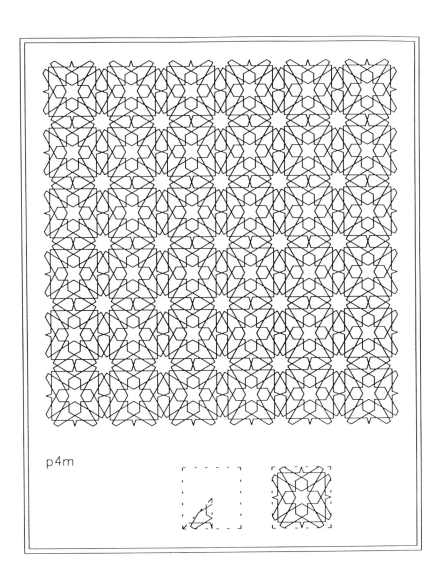

p4m

p4m

p4m

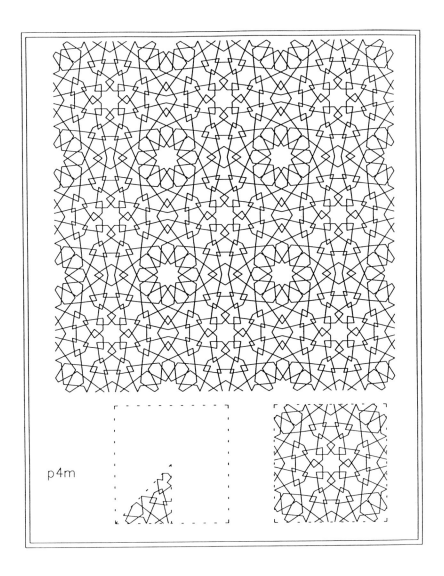

p4m

p4m

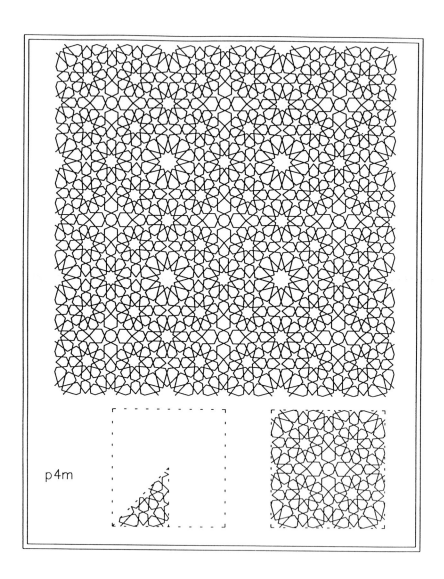

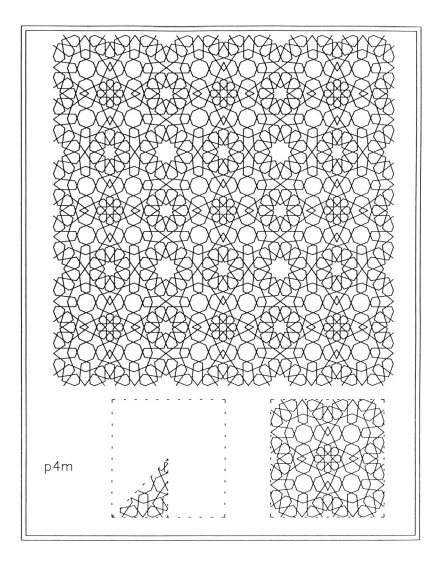
p4m

p4m

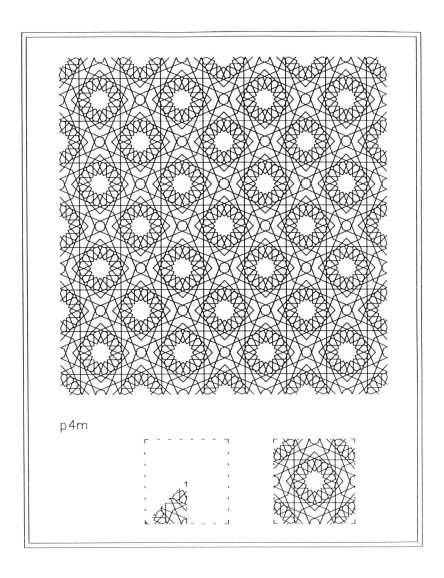

p4m

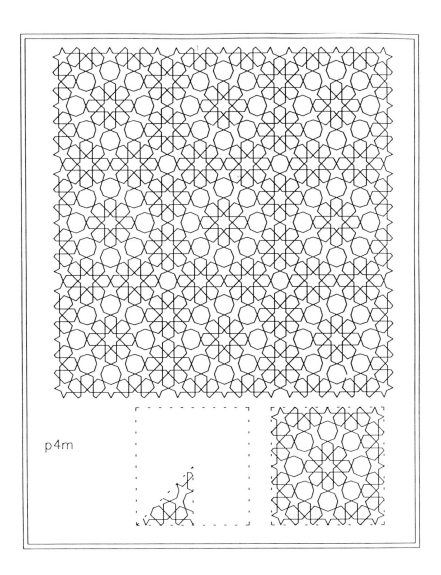

p4m

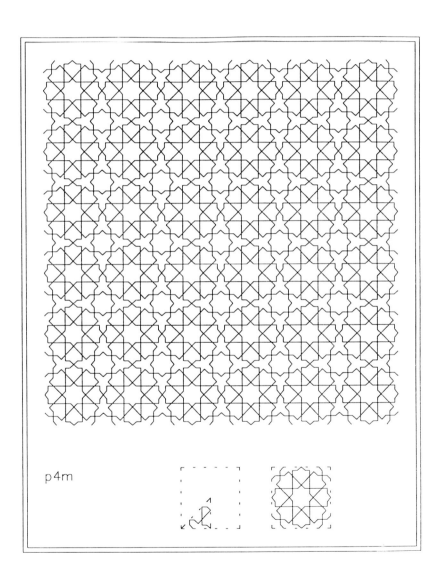

p4m

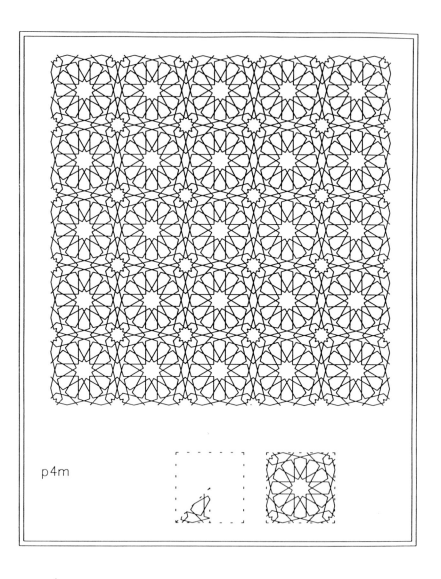

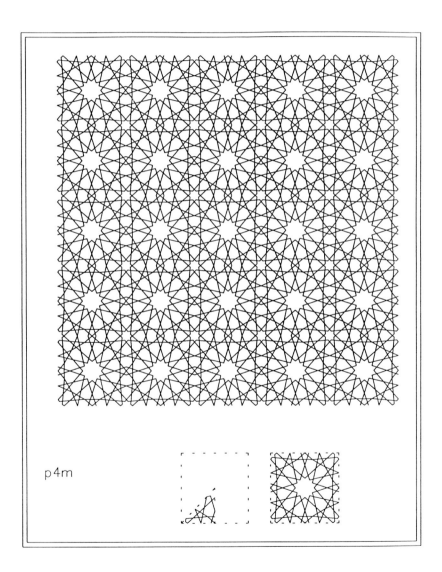

p4m

p4m

p4m

p4m

p4m

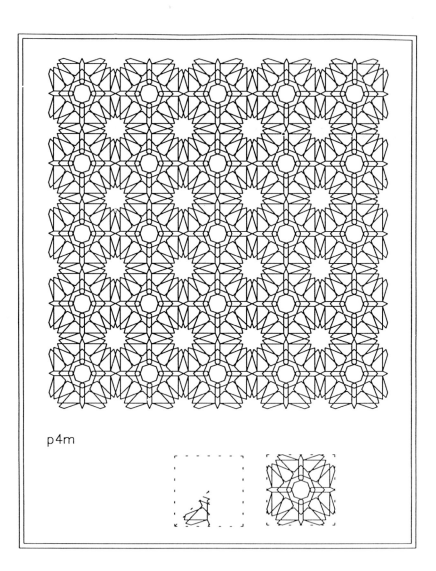

p4m

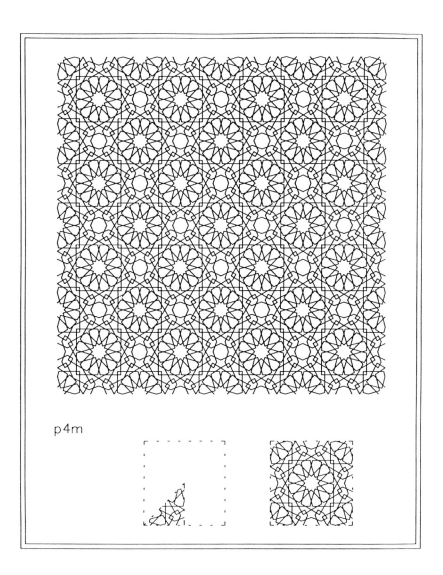

p4m

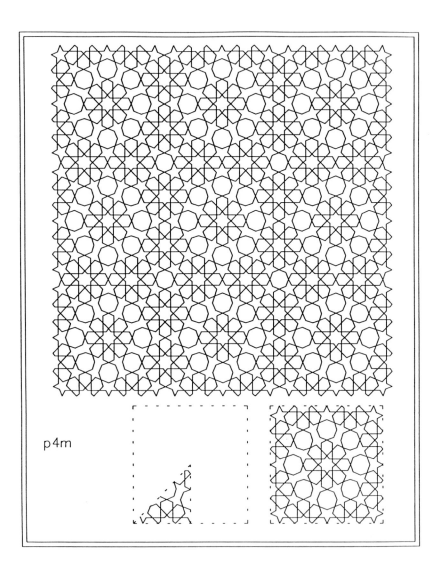

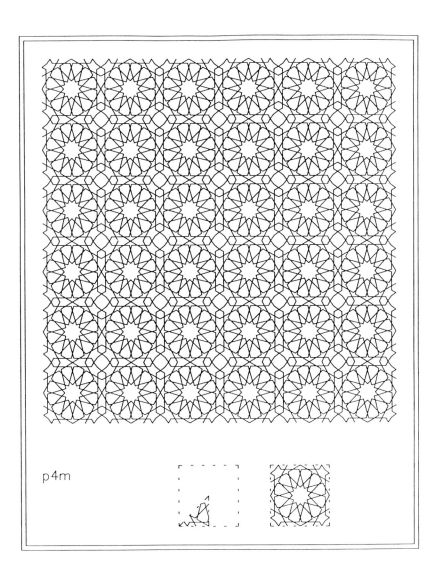

p4m

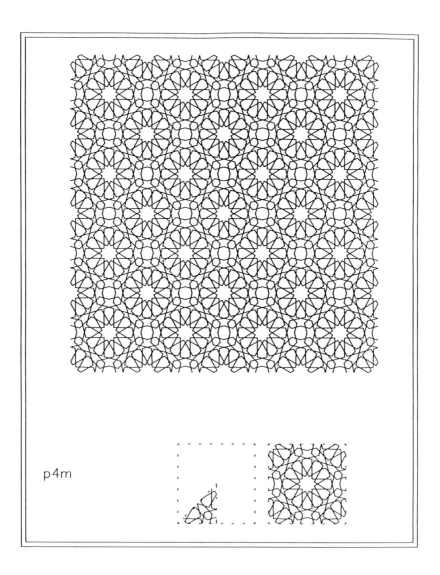

p4m

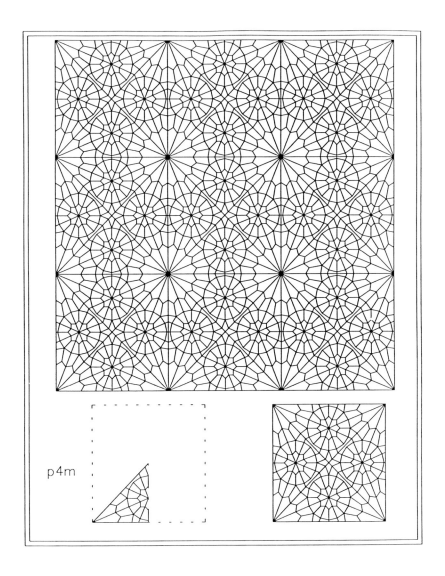

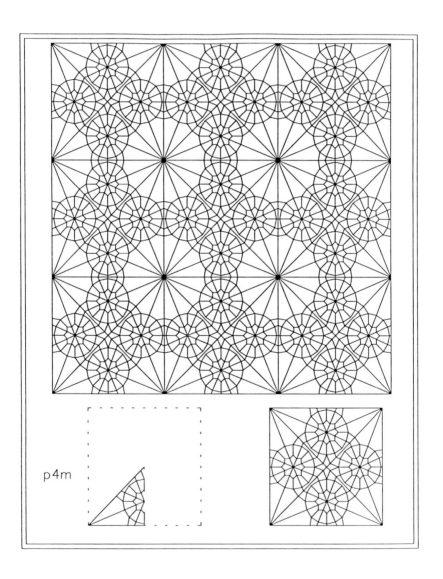

p4m

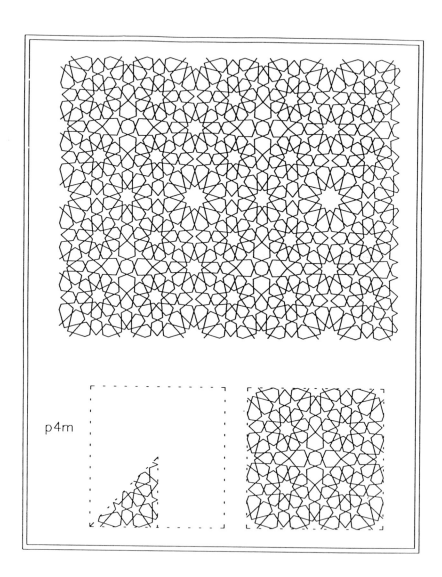

289

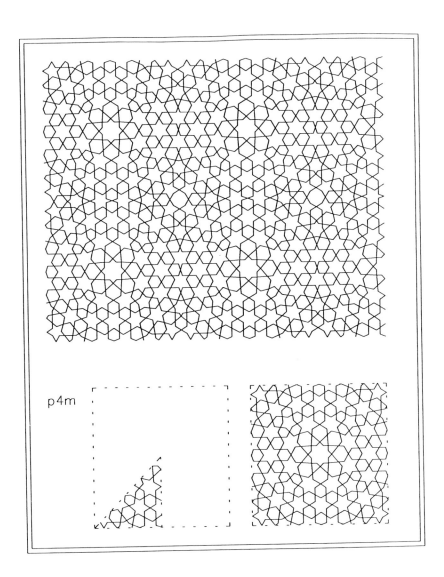

p4m

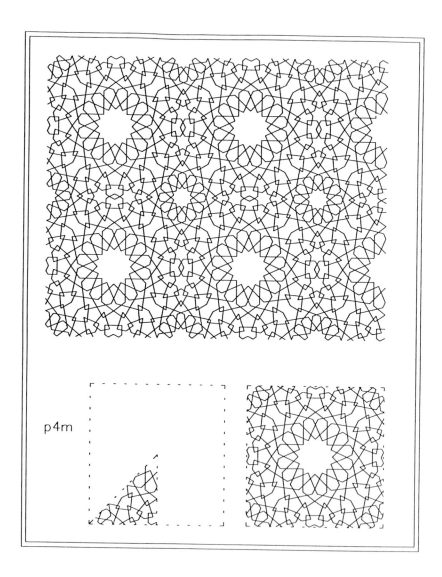

p4m

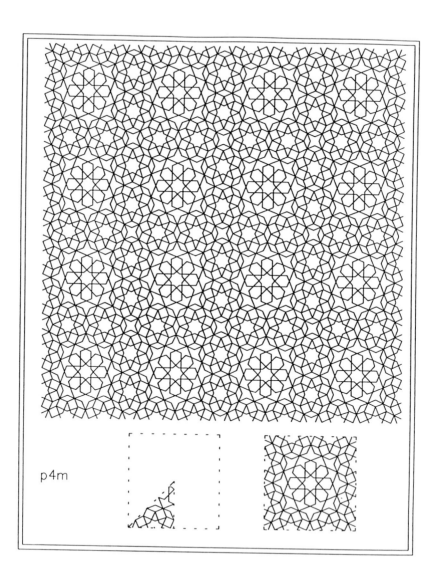

p4m

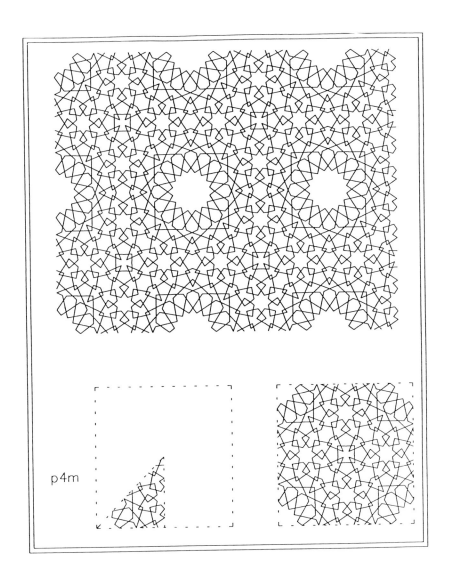

p4m

p4m

p4m

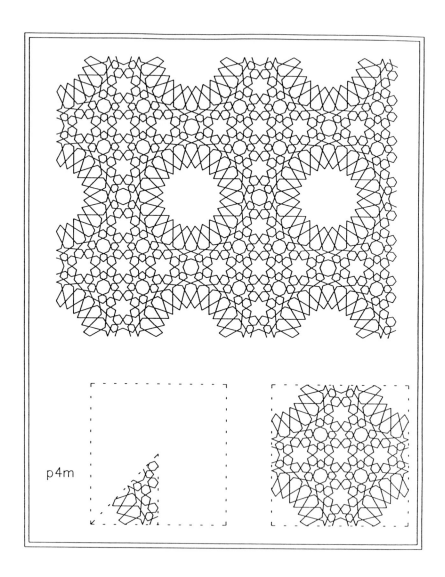

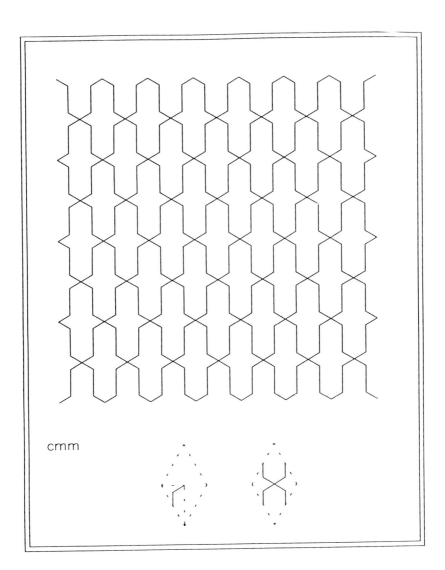

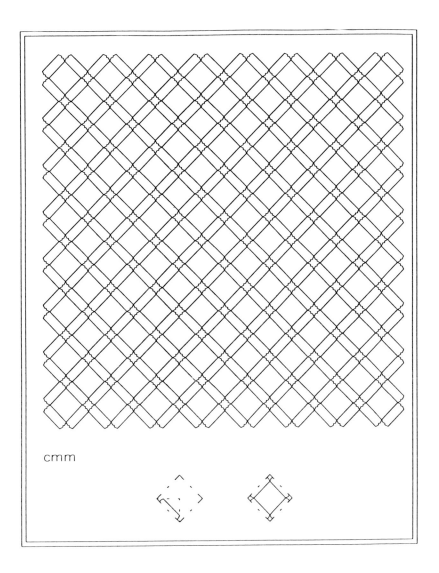

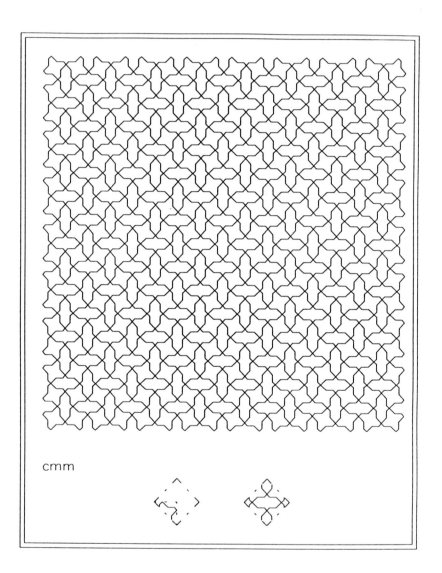

cmm

cmm

cmm

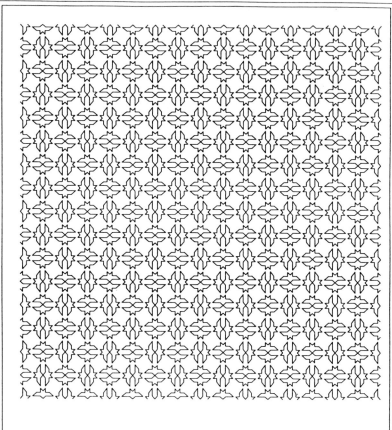

cmm

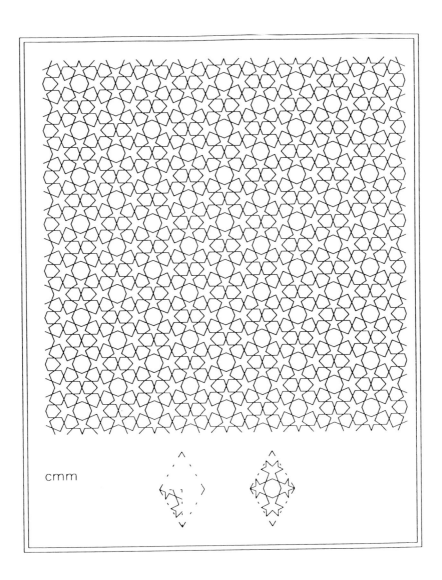

cmm

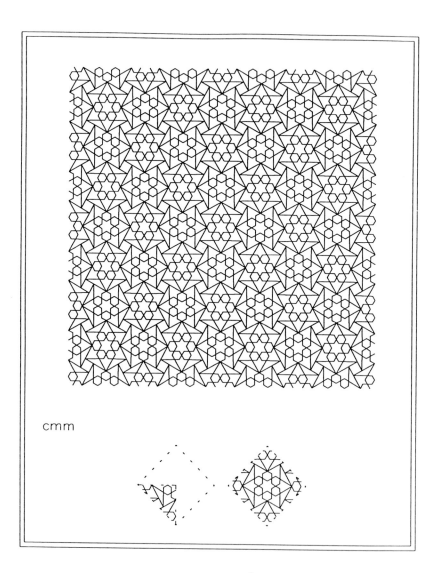

cmm

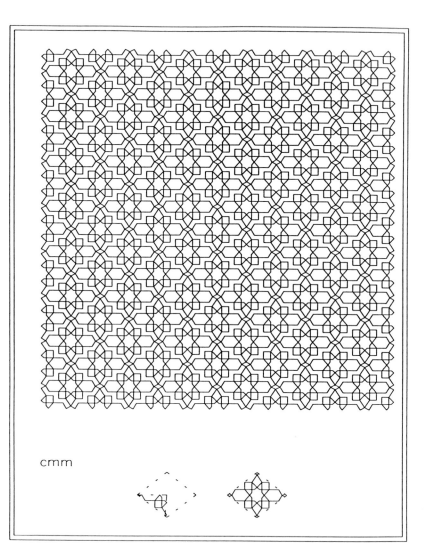

cmm

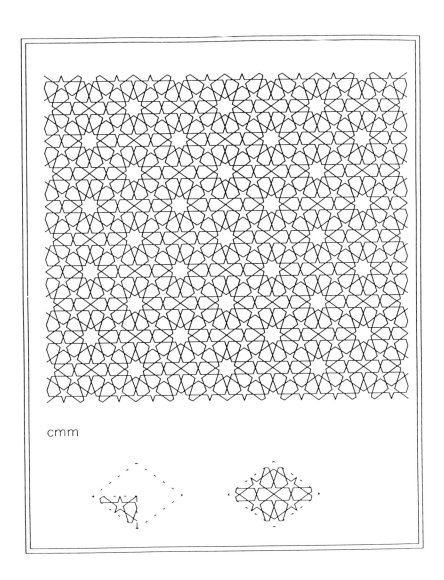

cmm

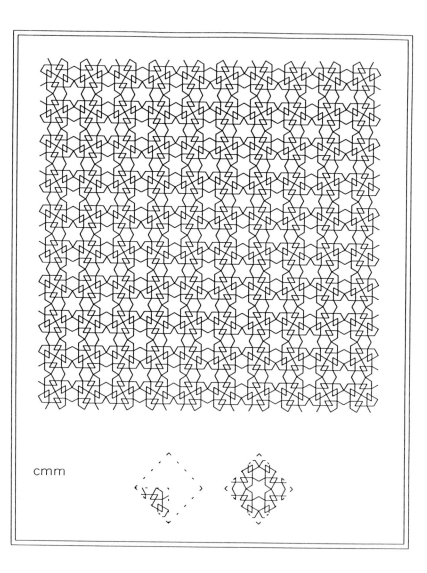

cmm

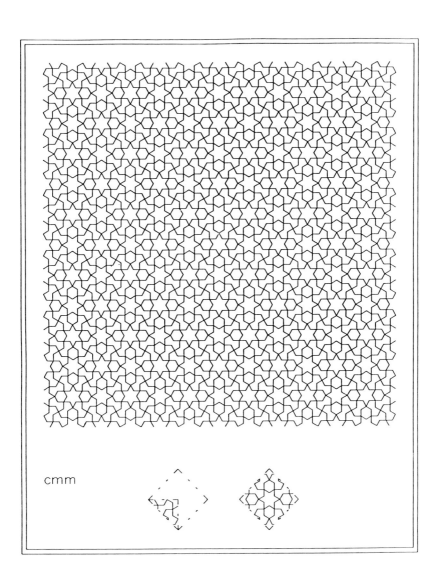

cmm

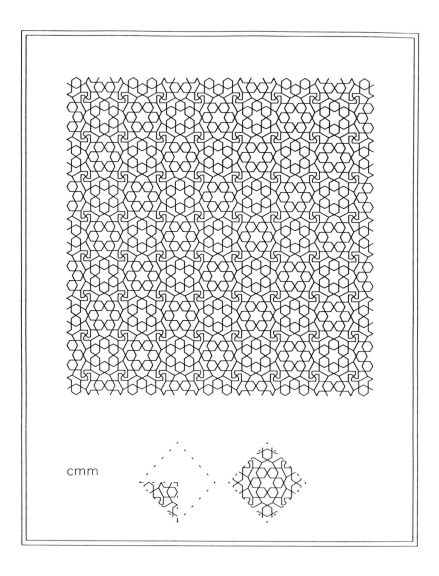

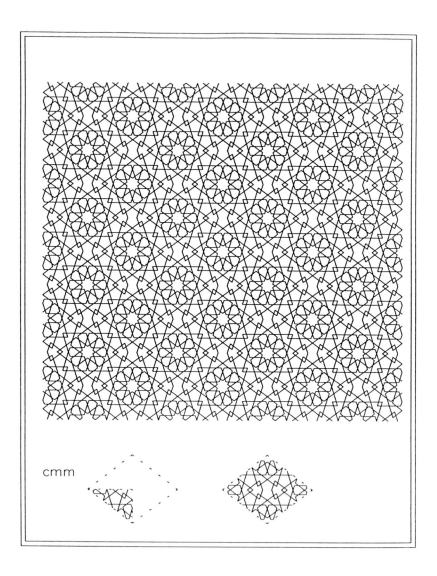

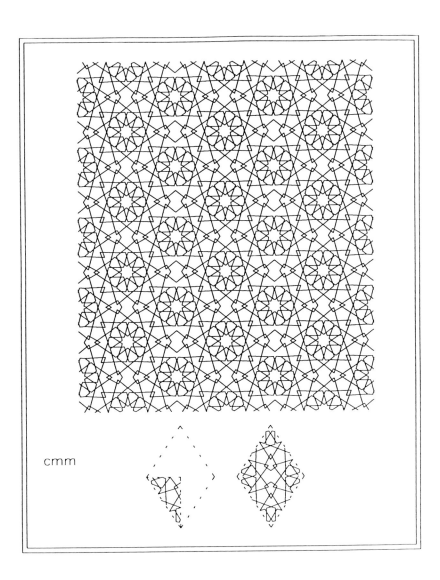

cmm

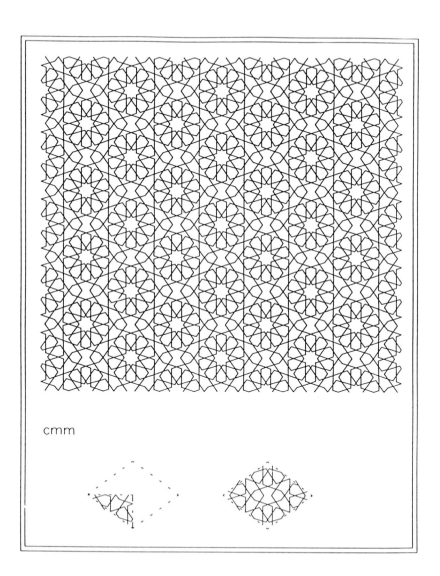

cmm

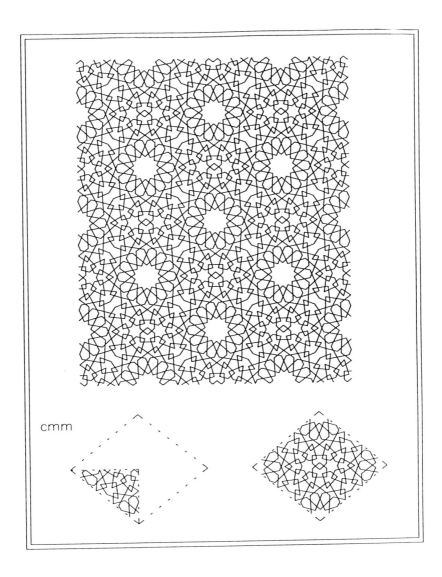

cmm

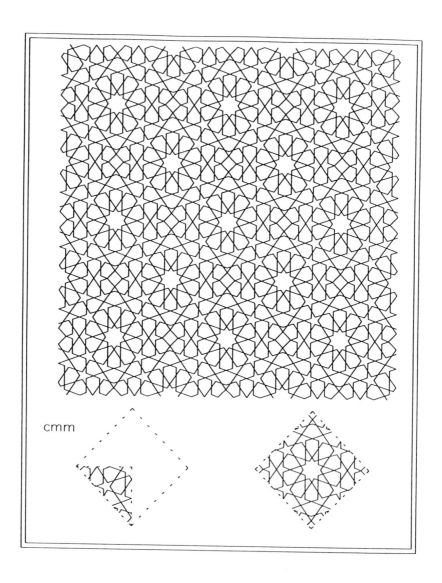

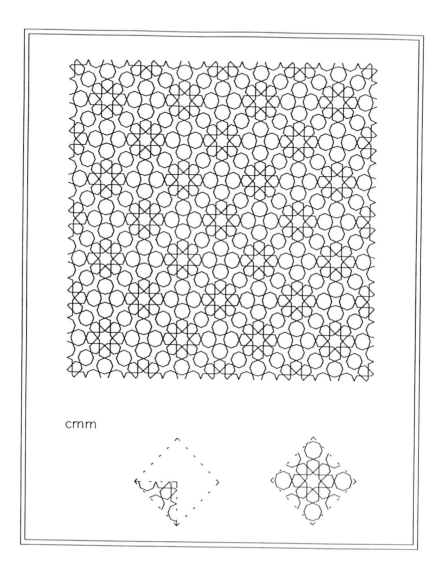

cmm

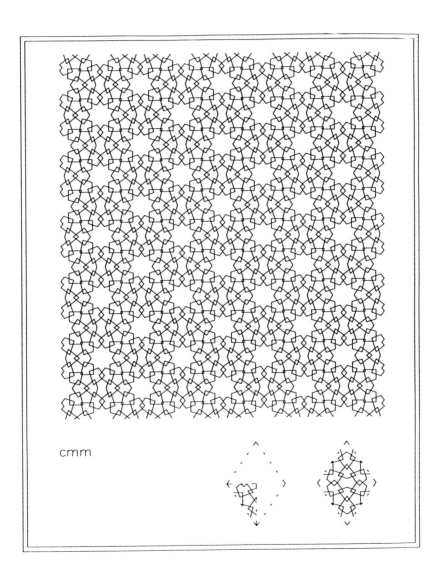

cmm

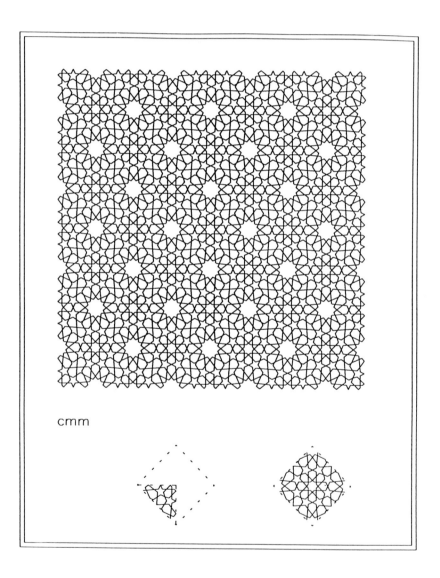

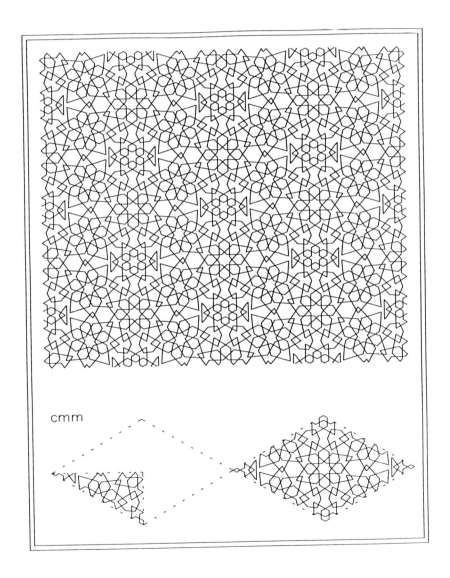

cmm

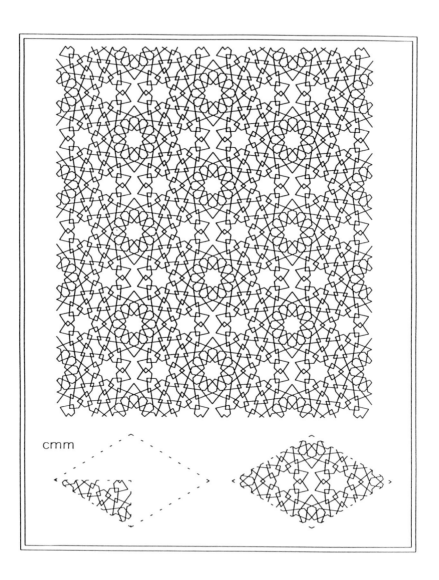

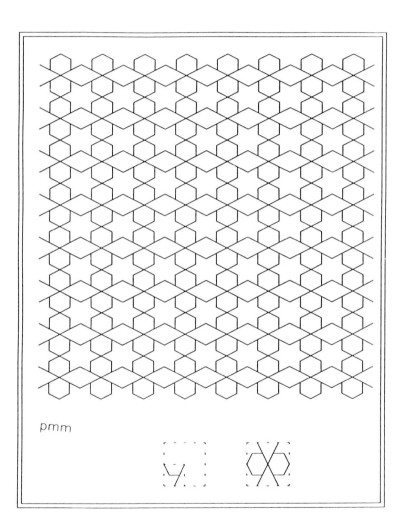

pmm

pmm

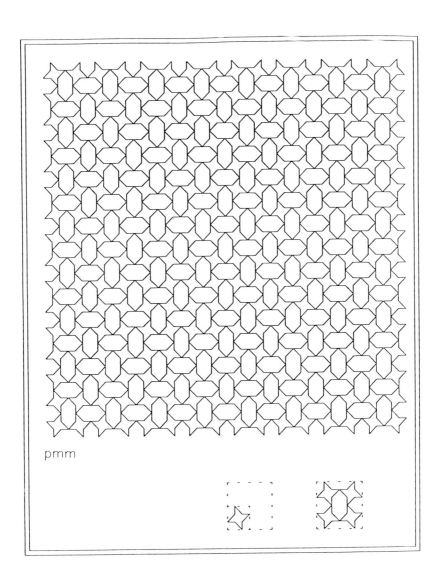

pmm

pmm

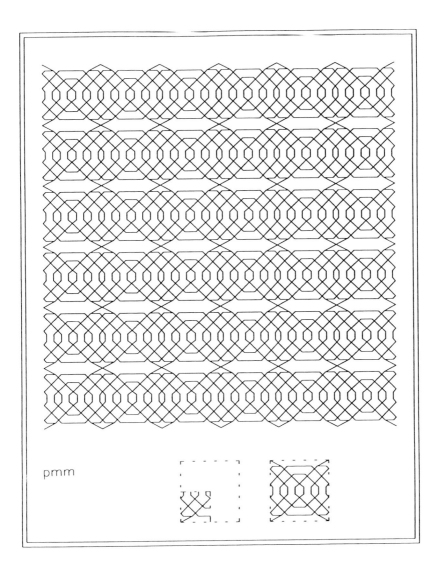

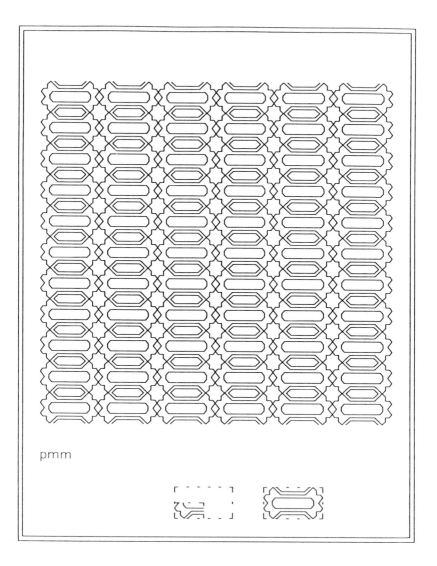

pmm

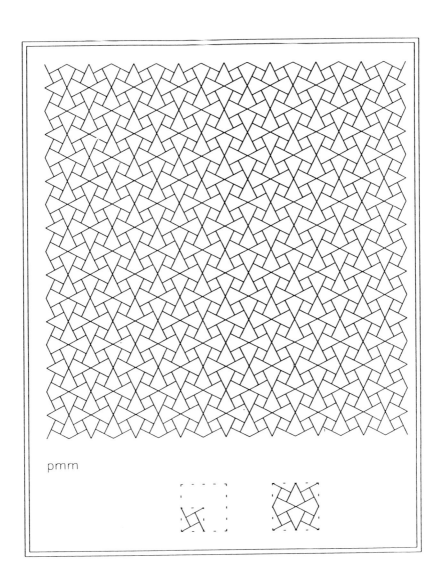

pmm

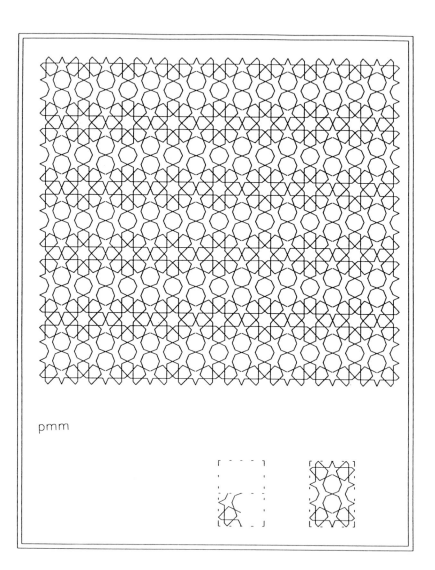

pmm

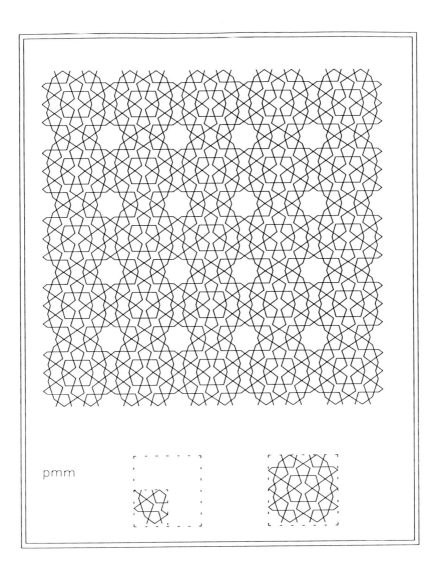

pmm

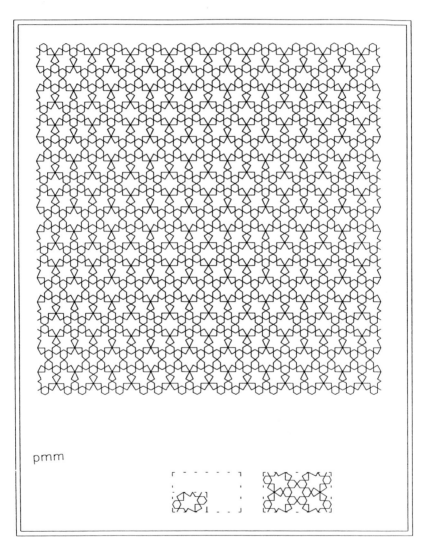

pmm

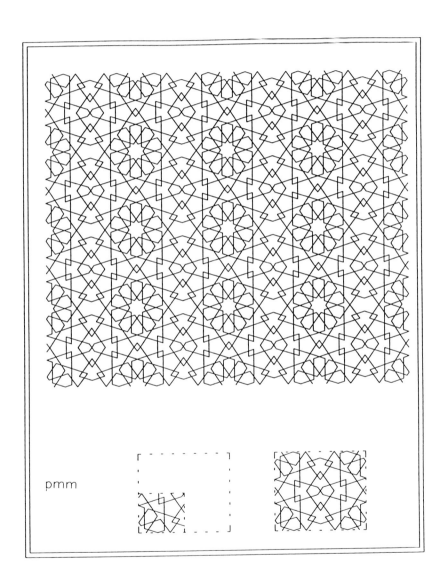

pmm

pmm

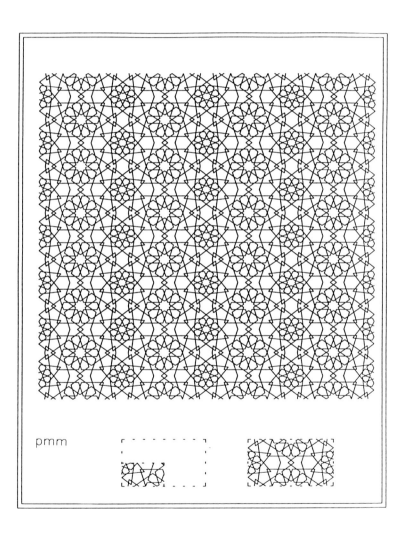

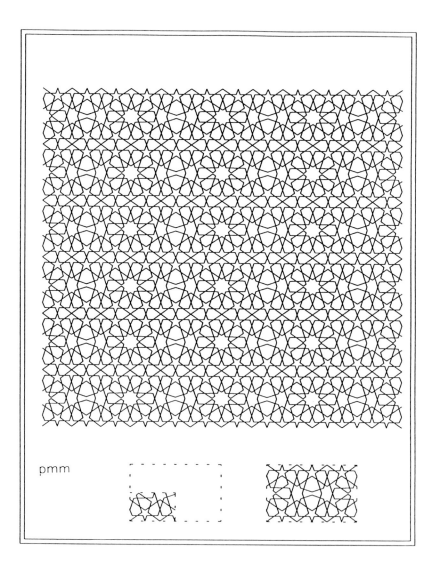
pmm

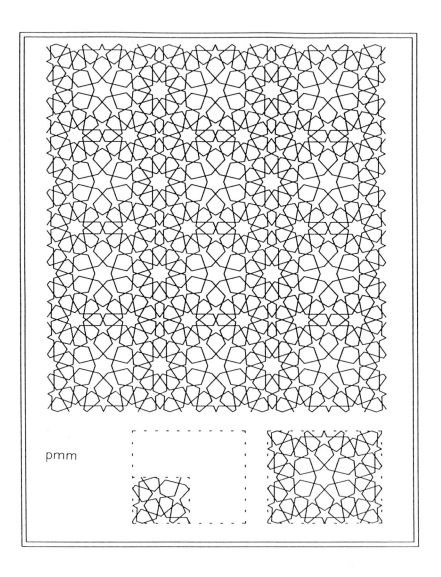

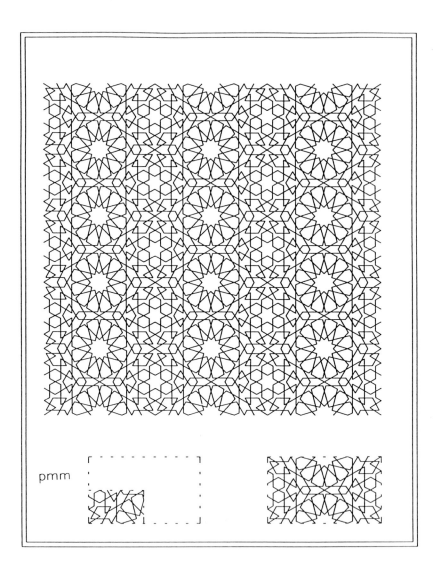

p6

p6

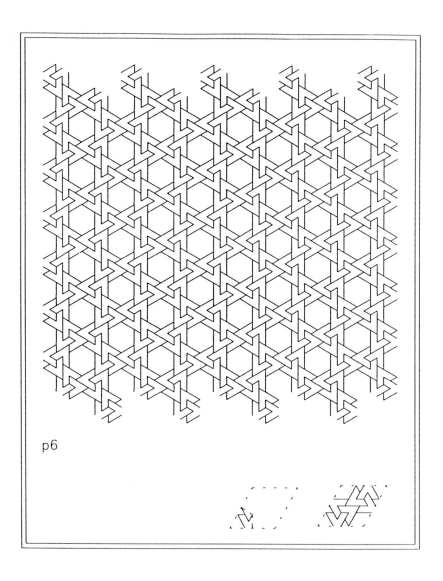

p6

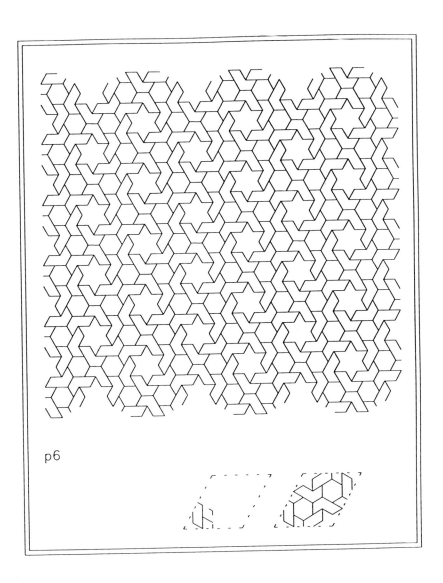

p6

p6

p6

p6

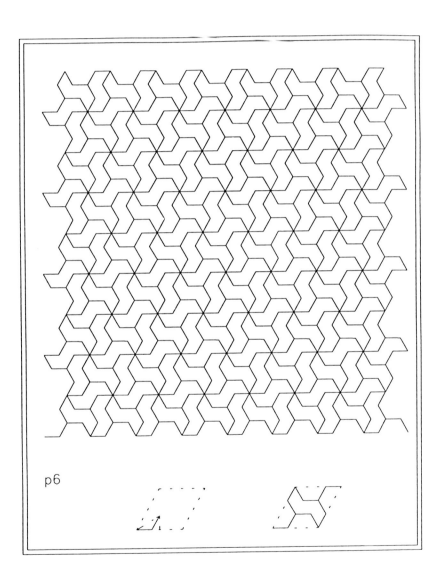

p6

p6

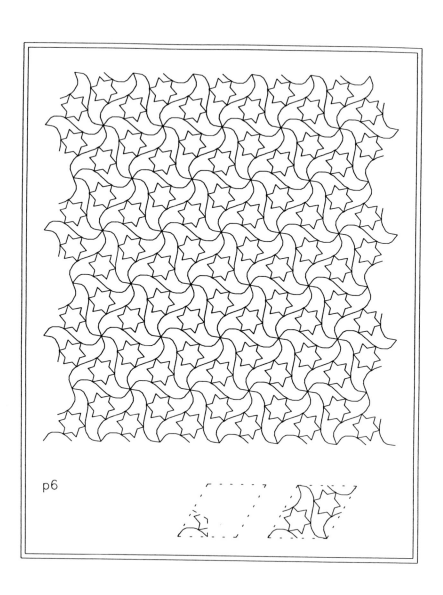

p6

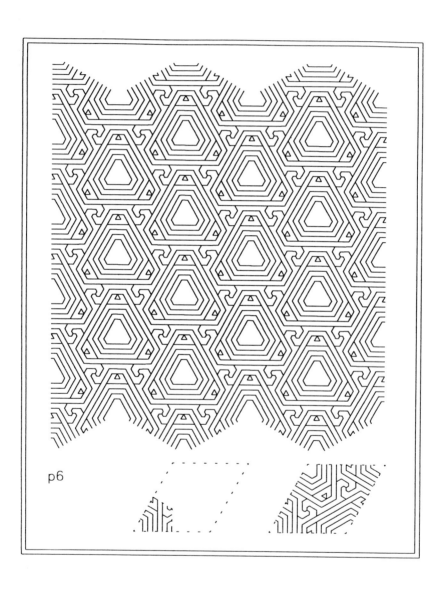

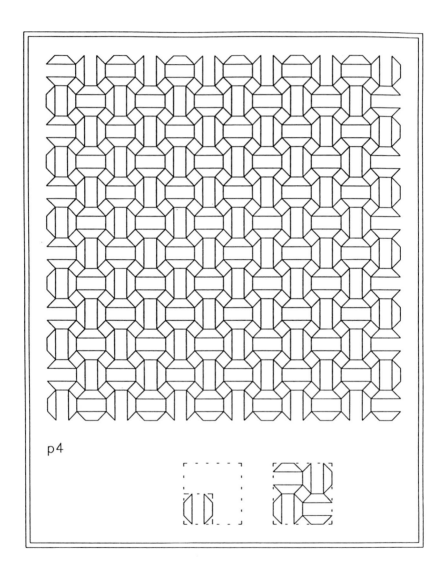

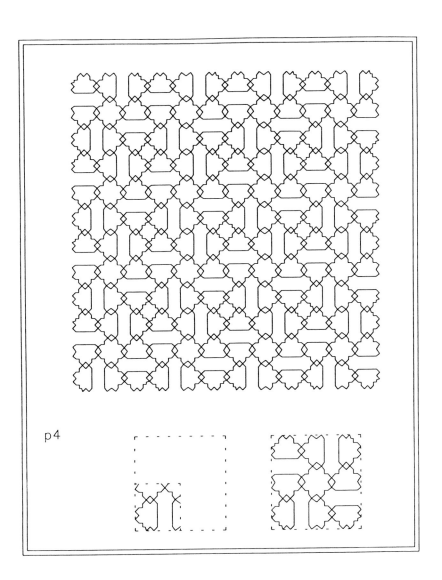

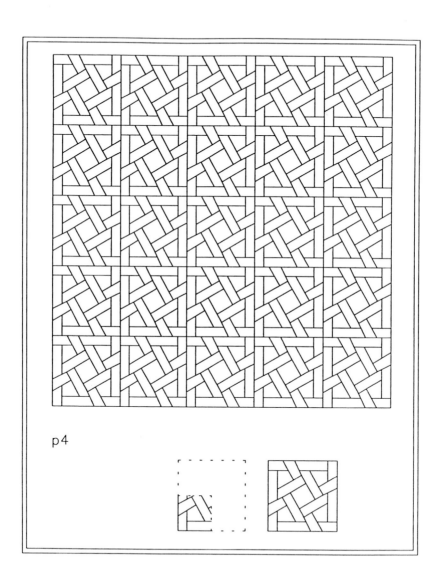

p4

367

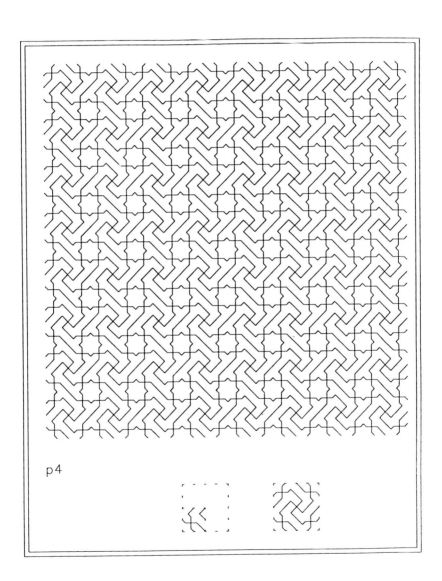

p4

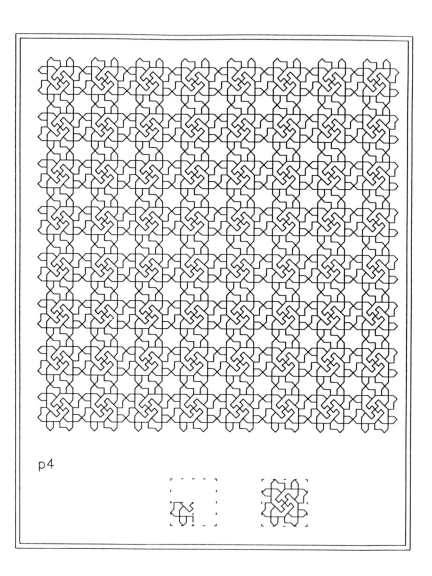

p31m

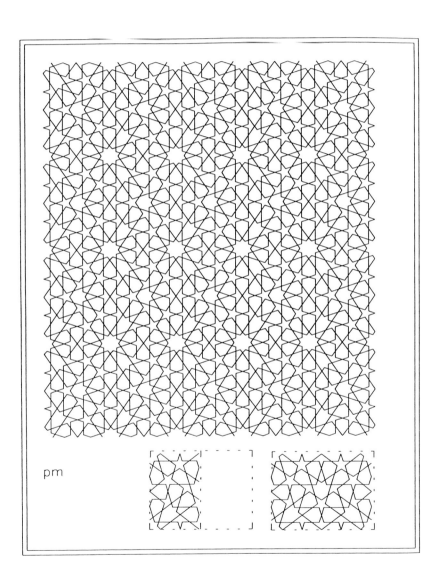

pm

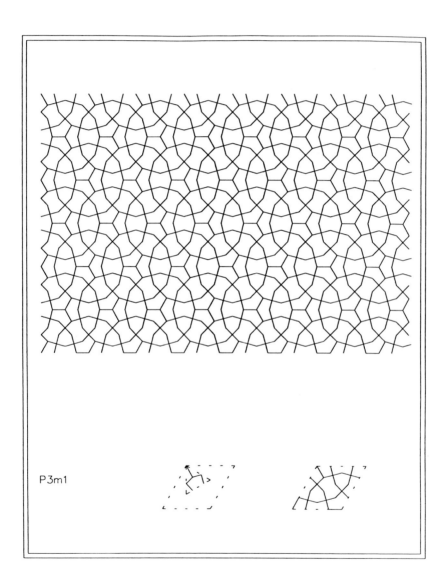

P3m1

377

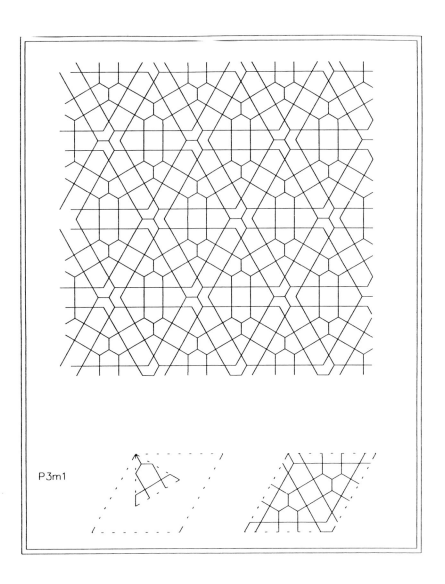

P3m1

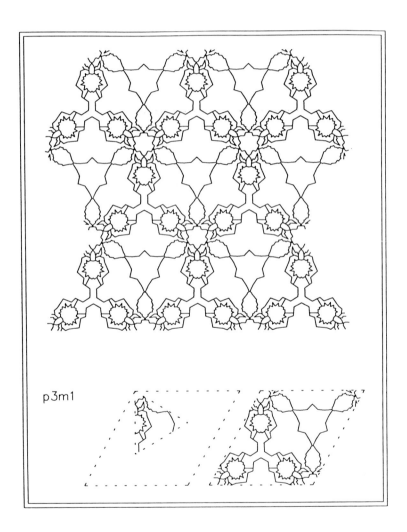

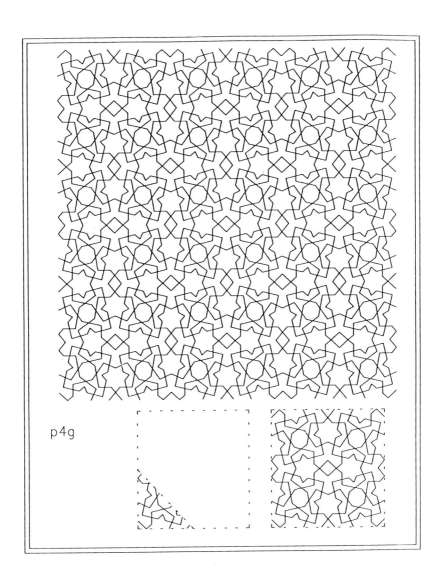

380

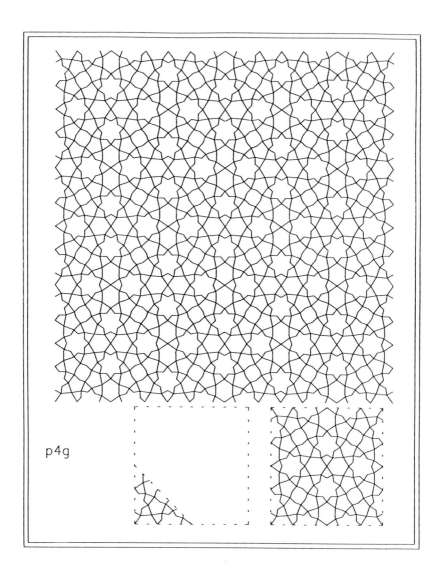

p3

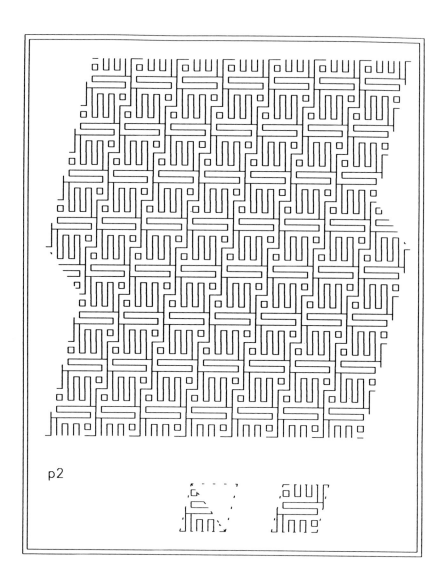

p2

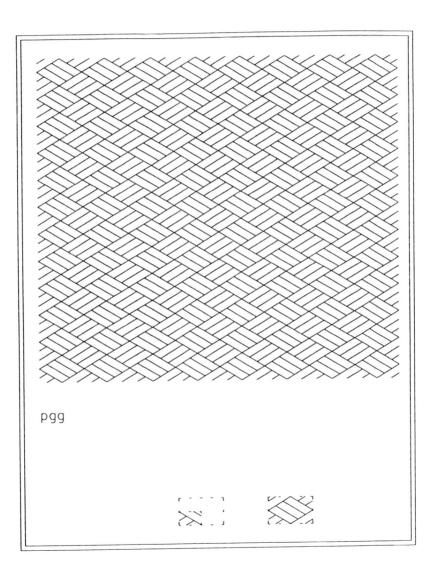

pgg

pg

REFERENCES

(1) Abas, S. J., *Computer Graphics Studies of Islamic Geometrical Patterns*, Proceeding of the Fourth International Conference on Computer Graphics, Published in Automatica, 31–2, pp. 11–24, (1990).

(2) Abas, S. J. and Salman, A. S, *Geometric and Group-theoretic Methods for Computer Graphic Studies of Islamic Symmetric Patterns*, Computer Graphics Forum, vol. 11, no. 1, pp. 43–53, (1992).

(3) Aslet, C., *Art is Here: The Islamic Perspective*, Leighton House, Country Life, vol. 16, 1642–1643, (1983).

(4) Bell, E. T., *Mathematics, Queen and Servant of Science*, New York, (1951).

(5) Bixler, N. H., *A Group Theoretic Analysis of Symmetry in Two-Dimensional Patterns from Islamic Art*, Ph.D. Thesis, New York University, (1980).

(6) Bourgoin, J., *Arabic geometrical pattern and design*, Firmin-Didot, Paris (1879), Dover, New York (1973).

(7) Bronowski, J., *The Ascent of Man*, British Broadcasting Corporation, London, (1973).

(8) Chorbachi, W. K., *In the Tower of Babel: Beyond Symmetry in Islamic Design*, Computer Math. Applic, vol. 17B, 751–789 (1989).

(9) Critchlow, K., *Islamic Patterns An Analytical and Cosmological Approach.*, Thames and Hudson, London (1976), paperback edition (1983).

(10) Crowe, D. W., *The Geometry of African art III: The Smoking Pipes of Begho*, in *The Geometric Vein (The Coxeter Festschrift)*, Springer-Verlag, New York, (1981).

(11) D'Avennes, P., *L'Art Arabe d'Après les Monument do Kaire depois le VIIe siècle jusqu'à la fin du XVIIIe siècle*, Vve.A. Morel et Cie., Paris (1869–1877). Reprint ed. *Arabic art in colour*, Dover Publications, New York, (1978).

(12) Dixon, R., *Geometry Comes Up To Date*, New Sientist, 5 May 1983.

(13) El-Said, I., and Parman, A., *Geometrical Concepts in Islamic Art*, World of Islam Festival Publ. Co., London (1976).

(14) Grünbaum, B., Grünbaum, Z. and Shephard, G. C., *Symmetry in Moorish and Other Ornaments*, Comp.& Math. with Appls. vol. 12B, 641–653, (1986).

(15) Grünbaum, B. and Shephard, G. C., *Tilings and Patterns*, Freeman, New York, (1987).

(16) Hankin, E. H., *Some Difficult Saracenic Designs*, Math. Gazette vol. 18, 165–168 (1934), and 20, 318–319 (1936).

(17) Hardy, G. H., *A Mathematician's Apology*, Cambridge University Press, Cambridge, (1977).

(18) Hargettai, I (ed), *Symmetry 1: Unifying Human Understanding*, Pergamon, Oxford, (1986) and *Symmetry 2*,(1989).

(19) Hoffmann, R., *How Should Chemists Think*, Scientific American, February 1993.

(20) Jeans, J., *The Mysterious Universe*, Cambridge University Press, Cambridge, (1937).

(21) Jones, O., *The Grammar of Ornament*, Day and Son, London (1856), recent reprint Studio Editions, London (1988).

(22) Keyser, C.J., *The Group Concept*, in *The World of mathematics*, George Allen and Unwin, London, (1960).

(23) Lalvani, H., *Coding and Generating Islamic Patterns*, Published by the National Institute of Design, Ahmedabad, India, (1982).

(24) Lalvani, H., *Coding and Generating Complex Periodic Patterns*, The Visual computer, 5, pp. 180–202, (1989).

(25) McGregor, J. and Watt, A., *The Art of Microcomputer Graphics*, Addison-Wesley, Wokingham(1984).

(26) Makovicky, E. and Makovicky, M., *Arabic Geometrical patterns — a treasury for crystallographic teaching*, t Jahrbook für Mineralogie Monatshefte, 2, 58–68, (1977).

(27) Makovicky, E., *Symmetrology of Art: Coloured and Generalized Symmetries*, Comp. Math. with Applic., vol. 12B, nos. 3/4, 949–980, (1986).

(28) Makovicky, E., *Ornamental Brickwork, Theoretical and Applied Symmetrology and Classification of Pattern*, Comp. Math. with Applic., vol. 17, nos. 4–6, pp. 955–999, (1989).

(29) Makovicky, E., *800-Year-Old Pentagonal Tiling from Marāgha, Iran, and the New Varieties of Aperiodic Tiling it Inspired*, in *Fivefold Symmetry*, World Scientific, Singapore, (1992).

(30) Mamedov, K. S., *Crystallographic Patterns*, Comp. Math. with Applic., vol. 12B, nos. 3/4, pp. 511–529, (1986).

(31) Martin, G. E., *Transformation Geometry. An Introduction to Symmetry*. Springer-Verlag, Berlin(1982).

(32) Montesinos, J. M., *Classical Tesselations and Three-Manifolds*, Springer-Verlag, New York, (1987).

(33) Muller, E.
  (a) *Gruppentheoretische und Strukturanalytische Untersuchungen der Maurischen Ornamente aus der Alhambra in Granada.* (Ph.D. Thesis, University of Zurich) Baublatt, Rüschlikon, (1944).
  (b) *El Estudio de Ornamentos como Applicacion de la Teoria de los Grupos de Orden Finito.* Euclides (Madrid) 6(1946) 42–52.

(34) Nasr, S. H., *Islamic Science an Illustrated Study*, World of Islam Festival Publishing Company, (1976).

(35) Niman, J. and Norman, J., *Mathematics and Islamic Art*, Amer. Math. Monthly vol. 85, 489–490, (1978).

(36) Paccard, A., *Traditional Islamic Craft in Moroccan Architecture*, Published by Editions ateliers 74, 74410 Saint-Jorioz, France, vols. 1 and 2, (1980).

(37) Palmer, E. H., *The Rubáiyát of Omar Khyám AND OTHER PERSIAN POEMS*, Edited by A. J. Arberry, Dent, London, (1976).

(38) Pérez-Gómez, R., *The Four Regular Mosaics Missing in the Alhambra*, Comp. Math. with Applic., vol. 14, nos. 2, pp. 133–137, (1987).

(39) Pérez-Gómez, R., et al., *La Alhambra*, an edition of *Epsilon*, published by Asociación de Profesores de Matemáticas de Andalucía, Granada, (1987).

(40) Polya, G., *Über die Analogie der Kristallsymmetrie in der Ebene*, Zeitschrift für Kristallographie, 60, 278–282 (1924).

(41) Rosen, J., *Symmetry at the Foundation of Science*, Computers & Mathematics, vol. 17, nos. 1–3, (1989).

(42) Salman, A., *Computer Graphics Studies of Islamic Geometrical Patterns and Designs*. Ph.D. Thesis University of Wales, Cardiff (1991).

(43) Schattschneider, D., *The Plane Symmetry Groups: Their recognition and notation*, The American Mathematical Monthly, 85, 439–450, (1978).

(44) Shubnikov, A. V. and Koptsik, V. A., *Symmetry in Science and Art*, Nauka, Moscow, (1972), Plenum Press, New York, (1974).

(45) Speiser, A., *Die Theorie der Gruppen von endlicher Ordnung*, Second Edn. Springer, Berlin (1927); Third Edn. Springer, Berlin (1937); Dover, New York (1943); Fourth Edn. Birkhauser, Basel (1956).

(46) Stewart, I. and Golubitsky, M., *Fearful Symmetry — Is God a Geometer?*, Blackwell Publishers, Oxford, (1992).

(47) Stryer, L., *Biochemistry*, W.H.Freeman, New York, (1988).

(48) Taft, K. L. and Lippard, S. J., *Synthesis and Structure of* $[Fe(OMe)_2(O_2CCH_2Cl)]_{10}$, *A Molecular Ferric Wheel*, Jour. Amer. Chem. Soc, vol. 112, pp. 9629–9630, (1990).

(49) Turnbull, H., *The Great Mathematicians*, Methuen & Co. Ltd., London, (1929).

(50) Wade, D., *Pattern in Islamic Art*, Cassell & Collier Macmillan, London (1976).

(51) Washburn, D. and Crowe, D., *Symmetries of Culture*, University of Washington Press, (1988).

(52) Weyl, H., *Symmetry*, Princeton University Press, Princeton, New Jersey(1952), Paperback reprint, (1982).

(53) *Zillïj: The Art of Moroccan Ceramics*, Eds. J. H. Hedgecoe and S. S. Damluji, Garnet Publishing Ltd., Reading, (1992).

INDEX

Arabesque   3
Associativity   60, 69

Basis   63

Centered cell   77
Characteristic shapes   8
Closure   68
Computer   31
Crystallographic Notation   77
Crystallographic restriction   63

Definition: Of Islamic Patterns   6
Dihedral group   59
DNA   31

Erlangen programme   70

Fundamental region   48, 79

Generator region   48, 79
Glide reflection   57
Ground symmetry   89
Group   68

Identity   58, 69
Invariant   57, 69
Inverse   61, 69
Islamic Ferric Wheel   41
Isometry   58

Khatem Sulemani   14

Kufic patterns 3

Lattice unit 48

Mendelyev Dmitri 31

Net types 75

Pattern 30
Pattern: 17 types 76
Primitive cell 77

Quran 71

Repeat Pattern 74
Repeat unit 48
Replication 39
Root Two System 18

Space filling 16
Symmetries of a Square 55
Symmetry 32, 57, 58, 67

Template motif 48, 79
Transformation 57
Translation 56

Unit cell 48, 74
Unit motif 48, 79
Unity 35

Wallpaper pattern 74

Zalij 24